Apocalyptic Conspiracism

ALSO AVAILABLE FROM BLOOMSBURY:

UFOs, Conspiracy Theories and the New Age, David G. Robertson
The New Apostolic Reformation, Trump, and Evangelical Politics,
Damon T. Berry
American Evangelicals and Muslims, Ashlee Quosigk

Apocalyptic Conspiracism

American Evangelicalism in an Age of Climate Crisis

TOM ALBRECHT AND TRISTAN STURM

BLOOMSBURY ACADEMIC
LONDON • NEW YORK • OXFORD • NEW DELHI • SYDNEY

BLOOMSBURY ACADEMIC
Bloomsbury Publishing Plc
50 Bedford Square, London, WC1B 3DP, UK
1385 Broadway, New York, NY 10018, USA
29 Earlsfort Terrace, Dublin 2, Ireland

BLOOMSBURY, BLOOMSBURY ACADEMIC and the Diana logo are trademarks of
Bloomsbury Publishing Plc

First published in Great Britain 2025

Copyright © Tom Albrecht and Tristan Sturm, 2025

Tom Albrecht and Tristan Sturm have asserted their rights under the Copyright,
Designs and Patents Act, 1988, to be identified as Authors of this work.

For legal purposes the Acknowledgments on p. viii constitute an extension of
this copyright page.

Cover design: Elena Durey
Cover image: Wooden post and sign © Sycikimagery/Getty

All rights reserved. No part of this publication may be reproduced or transmitted in any
form or by any means, electronic or mechanical, including photocopying, recording,
or any information storage or retrieval system, without prior permission in writing
from the publishers.

Bloomsbury Publishing Plc does not have any control over, or responsibility for, any third-party websites referred to or in this book. All internet addresses given in this book were correct at the time of going to press. The author and publisher regret any inconvenience caused if addresses have changed or sites have ceased to exist, but can accept no responsibility for any such changes.

A catalogue record for this book is available from the British Library.

A catalogue record for this book is available from the Library of Congress.

ISBN: HB: 978-1-3504-4294-8
PB: 978-1-3504-4293-1
ePDF: 978-1-3504-4295-5
eBook: 978-1-3504-4296-2

Typeset by Deanta Global Publishing Services, Chennai, India
Printed and bound in Great Britain

To find out more about our authors and books visit www.bloomsbury.com and
sign up for our newsletters

For my mother, Iris, and my brother, Tim. —TA
For Iris. —TS

Contents

Acknowledgments viii

1 Introduction: Apocalyptic Conspiracism as a Way of Comprehending Crises 1

2 Social Epistemology, Power, and Climatic Counter-Knowledge 21

3 Analyzing Digital Knowledge Discourses/Spaces 43

4 American Evangelicalism, the Environment, and Apocalyptic Conspiracism 59

5 The Construction of Evangelical Apocalyptic Conspiracist Climate Change Discourses 85

6 Generation of Evangelical Apocalyptic Climate Counter-Knowledge 121

7 Climate Change and Evangelical Apocalyptic Geopolitics 137

8 Conclusion: Apocalyptic Friends and the Truth of the End 155

Notes 163
References 166
Index 192

Acknowledgments

I would like to thank my coauthor, Tristan Sturm, who originally motivated me to undertake this research several years ago and who has now contributed to the publication of this book. I would also like to thank Prof. David Livingstone, Prof. Crawford Gribben, and Prof. Mike Hulme for their helpful comments and suggestions at various stages of this research project. Thanks to all my friends for all the encouragement and distractions. I'd also like to thank my family—Iris, Tim, Grobi, Jojo, Leonie, Sami, Hanna, and especially Jonah, for their support. —Tom Albrecht

Thank you to all of my supportive friends and colleagues in geography and across the university at QUB, and at the Centre for Apocalyptic and Post-Apocalyptic Studies at the University of Heidelberg. And thank you to my families—those I was born to and the ones love made—Mom and Dad, Newt and Liz-Erin. —Tristan Sturm

1

Introduction

Apocalyptic Conspiracism as a Way of Comprehending Crises

Perhaps no present has been more future-oriented than our own, and much of this future-oriented thought is characterized by cataclysm. The present age often seems to be driven by images of its own End. In particular, the already perceptible ramifications of climate change and scientific predictions of a warming planet can suggest that world-changing disasters, mass extinctions, dystopian futures, and other catastrophic scenarios seem to approach what can be described as *apocalypses*. By apocalypse, we mean both religious and popular secular meanings, both revelation of knowledge or a transforming event "that reveals or makes manifest a vision of ultimate destiny" (O'Leary, 1998: 5) or "an event involving great destruction" (Concise Oxford English Dictionary, 2011). The term's more pessimistic meanings governs this book's conceptual understanding of apocalypse.

The imagination of climate change and environmental degradation as a world-changing global crisis (*eco-apocalypticism*) increasingly figures in apocalyptic imaginations. It is common for environmental activists to employ language that compares global warming, climate change, and natural disasters to biblical apocalyptic stories. Well-known examples include climate activist Greta Thunberg stating in 2019 at the *World Economic Forum* (WEF) meeting in Davos that "our house is on fire" (in a talk called "Averting a Climate Apocalypse"), Al Gore claiming that "every night on the TV news is like a nature hike through the book of Revelation," and the Doomsday Clock of the Bulletin of Atomic Scientists, which now includes climate change in their calculations, announcing in 2023 that humanity is ninety seconds to midnight, the closest to the End we have ever been. Scientifically derived predictions of droughts, increasing power and frequency of hurricanes and tornadoes, rising sea levels, mass extinction, and even possible global wars over natural resources can be easily transposed onto religious and non-religious beliefs about the End-Times (Veldman, 2019; Skrimshire, 2019, 2014; Sturm and Lustig, 2022). We are at a point in time where such linkages are now so common that "apocalyptic imaginaries . . . infuse the climate change debate" (Swyngedouw, 2010: 216).

Such transpositional imaginaries are not solely driven by images of our End in, for example, climate disaster movies such as *The Day After Tomorrow* (2004), *Interstellar* (2014), and *Mad Max: Fury Road* (2015). Rather, "environmental discourses have long been clothed in the language of Apocalypse" (Hulme, 2009: 345). Starting with publications like *The Population Bomb* (Paul Ehrlich, 1968) and *Limits to Growth* (Club of Rome, 1972), researchers, activists, and political organizations have been predicting the collapse of human civilization caused by an exploitation of the natural environment and an ongoing increase of the global population. Such environmentally determinist narratives teach a "causal primacy to environmental resources—and especially to presumed resource scarcities—in generating societal stress, breakdown and conflict" (Selby and Hoffmann, 2014: 748).

Still, not all End-Times discourses on climate change necessarily follow a Malthusian logic. Scientific warnings which predict that the expected 1.5°C increase in global temperature "would cause unavoidable increases in multiple climate hazards and present multiple risks to ecosystems and humans" (IPCC, 2022: 15) convey that a world-changing disaster is imminent. Although the use of apocalyptic language in climate change debates is also criticized as an unscientific, rather populist, and post-political promotion of fear and resolve (Swyngedouw, 2010), the imagination of the climate crises as an apocalypse profoundly influences public, scientific, and political representations of climate change.

Yet, apocalyptic framings of climate change are versatile and therefore deviate from the environmentalist logic outlined above. In this book, we engage with parts of a highly complex (Margolis, 2020; Worthen, 2014; Swartz, 2012, 2011), powerful (Schwadel, 2017; Miller, 2014; Brint and Abrutyn, 2010; Durbin, 2018), and rightward resonant (Connolly, 2008; Barkun, 1996; Dochuk, 2011; Berry, 2017; O'Donnell, 2022; Sturm and Lustig, 2022; Bloom and Rollings, 2022; Dalleck, 2023) movement whose perceptions of public political issues like climate change and the Covid pandemic have a pervasive impact on American public debates (Crossley, 2021; Alder and Schäublin, 2020; Jenkins et al., 2018): American evangelical apocalypticism.

While American evangelicalism is a large and diverse Protestant Christian movement, we engage with predominantly white American evangelicals who emphasize End-Times theology, refer to the Bible as the utmost source of knowledge, and advocate religious and political conservatism as well as Christian nationalism and anti-globalism. For these evangelical Christians, the biblical apocalypse implies individual salvation and the return of the Messiah, Jesus Christ, to Earth. Here, a moral, ecological, social, and political decline is often welcomed as it anticipates the imminence of these desirable End-Times events (Sutton, 2017; Curry-Roper, 1990). But what is considered a felicitous sign of the times is contingent on preexisting religious, cultural, and also political beliefs. Some American evangelical apocalyptic Christians believe that the physical indicators of climate change, global warming, and environmental degradation signal the End-Times. But for other American evangelical Christians who await the imminent apocalypse, the disastrous developments of the past years are neither the decay of planet Earth nor the pandemic as a biblical plague. Rather, it is the political response to these crises that indicates that the end of time is near.

Driven by an apocalyptic geopolitical imaginary that expects the establishment of an anti-Christian world government over the course of the biblical End-Times, some American apocalyptic evangelicals perceive restrictive crisis politics in relation to climate change as indicators of a global dictatorship. Global climate protection policies are interpreted as part of a biblically announced deception that will bring the Antichrist into power. They argue that human-made global warming is an instrumental hoax which facilitates the population into accepting an authoritative global government, controlled by the Antichrist, as the solution for the non-existent climate crisis. Similar beliefs proclaim that the Covid pandemic is yet another fake crisis that is utilized to justify freedom constraints and surveillance measures that prepare the world for the Antichrist's global dictatorship. Such beliefs that emphasize crisis politics rather than the crisis itself connect historically established evangelical End-Times doctrine or eschatology to contemporary global developments to reconfirm the truth and infallibility of prophecy.

These apocalyptic perceptions of global crises are not just politically motivated; they are also conspiracist. Conspiracies and extensive global deceptions—which are developed and acted upon by politicians, scientists, and the mainstream media—are the means through which the end of contemporary societies is imagined by virtue of the establishment of a one-world government. We call this assemblage of thought *apocalyptic conspiracism* to conceptualize discourses that claim global crises such as climate change are conspiracies designed to actualize apocalyptic transformation.

To develop this linkage further, it is not just religious movements in the United States which conflate political responses to climate change as apocalyptic. Regardless of contrasting imaginations of the source of the evil that strives for a transformation to the worse, religious and secular apocalyptic discourses share and trade core arguments, as well as co-constitutionally reinforce and legitimize each other through discursive citation (DiTommaso, 2020; Barkun, 2013). More than this, white Christian nationalism motivates and maintains political racist and patriarchal orders that have deep roots in US history. Because of this, both (rightward) secular and evangelical discourses potentiate each other and are, in some ways, inseparable. Because of this, we argue that it is important to approach religious apocalypticism not as separated from non-religious fields of society but rather as shaped by cultural, social, scientific, and (geo)political histories (Rey, 2014; Taylor, 2007; Asad, 2003, 1993). This convergence can have alarming, immanent results, as illustrated by the storming of the American Capitol Building on January 6, 2021, where Christian and secular conspiracists and apocalypticists of the American right joined forces through resonant and shared perceptions of what we call apocalyptic conspiracism. They were driven by similar political motives, most importantly the opposition to the expected loss of freedom and the defense of the national sovereignty of the United States that is threatened by the prospect of a one-world government (Armaly, Buckley, and Enders, 2022).

Fields of Investigation

The guiding objective of this book is to explore the construction of apocalyptic conspiracist discourses that reject mainstream explanatory models of climate change, specifically knowledge-claims disseminated by the Intergovernmental Panel on Climate Change (IPCC, in the case climate change). In the most basic sense, these mainstream explanatory models state that anthropogenic climate change is a real issue which requires pervasive and coordinated political responses on a global scale to mitigate these crises.

We ask which discursive practices are used to delegitimize dominant mainstream knowledge and, subsequently, how alternative counter-knowledge on climate change is created, communicated, and justified. In that context, this book further examines how the authority of apocalyptic conspiracist knowledge-claims is developed and sustained while analyzing the underlying social, cultural, and (geo)political histories that constitute the foundation of contemporary powerful and influential conspiracist End-Times discourses.

This book focuses on the analysis of apocalyptic conspiracist discourses of climate change that originate from digital spaces connected to American evangelicalism. By apocalyptic conspiracist discourses, we mean texts that are representative of "intellectual constructs" (Barkun, 2016: 4) that postulate an approaching transformation of worsening conditions on a global scale and that are willingly initiated by evil superpower(s) who scheme to deceive the public for their respective benefit.

It is important to make clear that the grouping "evangelical" can connote contradicting religious, eschatological, and cultural attitudes. We therefore understand it as a heterogeneous movement. Due to several debates connected to the label "evangelical" (Margolis, 2020; Kidd, 2016; Worthen, 2014) and to avoid any possible generalizations, it is important to stress that this book does not attempt to provide an overarching analysis of the complexity and diversity of American evangelicalism, nor do we engage in mapping out its various eschatological belief systems. Evangelical Christianity no longer fits neatly into the tripartite pre-, post-, and a-millennialism categorizations. Rather, we observe an often messy eschatological landscape within evangelical Christianity, one shot through with influences from myriad theological sources, not least Pentecostalism and related Charismatic movements as well as Judaism. Indeed, we challenge this tripartite distinction at its core by exploring the networked narrative flexibility between various evangelical groups, the media, conspiracy theorists, and rightward groupings that pull from each other to build networked narratives of truth. Moreover, we recognize the essentialism of bounding analysis around "dispensational premillennialism" (in short, expectations of the rise of the Antichrist, Jesus' Second Coming, and the Rapture) as a coherent group given their networked narratives and ideologies around nationalism and civil religion. Our discussion of dispensational premillennialism, introduced below, is used both as a shorthand for a wide set of theological and ideological beliefs and as a genealogical exploration of the roots of "apocalyptic conspiracism." In what follows, we refer to this groupness as "premillennialists," "white apocalyptic evangelicals," or "evangelical apocalyptic conspiracists" when such conceptual descriptors are appropriate. For example, while all of the evangelical apocalyptic conspiracists we engage with are dispensational premillennialist at their core, not all dispensational premillennialists are conspiracists.

Methodologically, this book explores texts that fall in the category of apocalyptic conspiracism and that draw from American evangelical apocalyptic culture. However, since religious discourses cannot be analyzed separately from the world they emerge from, this research further considers rather (post)secular American conspiracist End-Times belief systems to examine the reciprocal relationship between evangelical apocalypticism and conspiracist discourses of the American right (Wilson, 2017; Barkun, 2013).

The specific sites of investigation and primary data collection are found in the digital world. Certainly, adherents of conspiracist belief systems do not merely communicate through the internet. Rather, ideas are also exchanged through face-to-face communication, through print media, by smaller niche radio and TV stations, and, importantly for the evangelical sphere, through sermons, resulting in "an alternative communications system cobbled together from a variety of both old and new media" (Barkun, 2013: 232).

Nevertheless, the internet profoundly changed the fields of apocalypticism and conspiracism and constitutes a pivotal space for the distribution of religious and secular conspiracist End-Times discourses (Wilson, 2017; Barkun, 2016; Howard, 2010, 2006). Due to the participatory features of the internet, reduced gatekeeping compared to traditional media platforms, and the possibility to find like-minded individuals outside one's social environment in the physical world, conspiracist and evangelical apocalyptic discourse can expand and thrive in the crucible of the internet (Barkun, 2013; Wilson, 2013). Therefore, we collected relevant primary data in what we approach as *digital spaces* to explore the particular knowledge-creating strategies and truth-claims of apocalyptic conspiracist online environments.

Engagement with some of the most popular evangelical End-Times websites revealed that it is important to approach digital apocalyptic conspiracism as an intellectual construct that gains clout and popularity through a mutually reinforcing citationality across a variety of websites. Texts on different websites stand in an intertextual relationship with each other in network-like structures of authors/websites cyclically citing each other and often republish the same texts (Hodges, 2015; Robertson, 2018; see Chapter 4). In short, through an in-depth textual analysis of publicly accessible digital resources published between 2015 and 2022, this book expands the academic literature on the discursive construction of anthropogenic climate change denialism, anti-globalist geopolitical imaginations, and the End-Times conspiracism of the American right.

In sum, this book argues that the cultures of American evangelical apocalyptic theology and right-wing conspiracism have formed a constituency—what we call apocalyptic conspiracism—that is productive of alternative truths concerning climate change in the crucible of digital spaces.

Apocalyptic Conspiracism

So far, the conceptual amalgam *apocalyptic conspiracism* has only been mentioned in passing in the academic literature (e.g., Sturm and Albrecht, 2021a; Wilson, 2017; Berlet, 2004), although we hold that it is probably the most suitable description of the fatalistic, fearmongering, and politically extremist conspiracy discourses which have become increasingly visible and powerful in recent years.

In the most basic sense, apocalyptic conspiracism describes the confluence of apocalyptic and conspiracist discourses to create a (more or less) coherent holistic belief system which postulates that the world will profoundly change for the worse as a result of a global network of interconnected conspiracies by evil and all-powerful cabals. Such End-Times theology supposedly exposes global conspiracies such as Agenda 21/30, the Great Reset, or the NWO, among other related geopolitical imaginations. Conspiracist End-Times beliefs claim to reveal the true nature of things to its adherents who perceive themselves as an elect minority which has exclusive access to the truth about the past, present, and future, and in particular, the hidden machinations of evil. Although apocalyptic conspiracism is primarily a pessimistic belief system that fetishizes looming doom and the inevitable flourishing of evil, it can also provide hope to adherents. Some discourses hold that the eventuality of such evil can be defeated and that individual or collective salvation can be reached. Still, apocalyptic disaster is a requirement for an age of prosperity where evil must fully rise before salvation is possible. Driven by their purported exclusive knowledge about the truth, apocalyptic conspiracists "believe to know" that a historical catastrophe is about to happen and, furthermore, that this imminent End is the result of conspiratorial endeavors. At the same time, apocalyptic conspiracist belief systems are flexible as they constantly adapt to the current state of the world to sustain their legitimacy and truth. Due to this constant adjustment to historical realities and sometimes absent End-Times catastrophes, dissonances within the belief system can emerge, but are most often found infelicitous by believers.

An important feature of the concept of apocalyptic conspiracism is that it explores not single *conspiracy theories* but rather belief systems that assume an interconnected network of global conspiracies (Barkun, 2013). In the most basic sense, a conspiracy theory can be defined as a "proposed explanation of events that cites as a main causal factor a small group of persons (the conspirators) acting in secret for their own benefit, against the common good" (Uscinski, Klofstad, and Atkinson, 2016: 58). Yet, many conspiracy beliefs today, in particular those with a religious origin, convey a holistic worldview in which almost every event or development of international importance can be incorporated and be related to each other as they are all part of a larger scheme that prepares the world for fundamental change (Asprem and Dyrendal, 2018; Wilson, 2017; Ward and Voas, 2011). Here, the term "conspiracism" describes "the belief that powerful, hidden, evil forces control human destinies" (Barkun, 2013: 2; cf. Muirhead and Rosenblum, 2019). Thereby, conspiracism does not mean single conspiracy theories, but what Robertson, Asprem, and Dyrendal (2018: 3) describe as the "full package deal," and, more specifically, "worldviews permeated by conspiracy beliefs and suspicious inferences, rather than to individual conspiracist narratives or beliefs" (2018). In other words, while conspiracy theories contradict selected official accounts of truth and argue that one or more single events are the result of a sinister plots, conspiracism describes an entire holistic worldview which holds that human history and destiny on a global scale is controlled by evil, all-powerful conspirators.

Conspiracism further addresses what Barkun (2013: 7) calls discourses of a *superconspiracy*, which "attribute all of the world's evil to the activities of a single plot, or set of plots." In conspiracism, virtually every major event, crisis, or development can be connected to each other as part of a superconspiracy that attempts to overthrow the current world order. For instance, climate change, Covid-19, the American presidential elections, or 5G networks can all be portrayed as individual plots that belong to the New World Order superconspiracy (Flaherty, Sturm, and Farries, 2022). This global magnitude and the focus on not some evil individual or group, but on global evil, makes conspiracism compatible with apocalypticism, as both discursive fields postulate "a hidden, overwhelming power that suffuses history, leaves traces for believers to find, and drives history towards a goal" (Robertson, Asprem, and Dyrendal, 2018: 3). Although conspiracist belief systems might appear complex to outsiders, "belief in a malevolent, virtually omnipotent conspiracy has the advantages, first, that is explains everything requiring explanation through a single factor; and, second, that all apparently disconfirming information may be dismissed as snares and delusions fabricated by the conspiracy itself" (Barkun, 1986: 152).

From Improvisational Millennialism to Apocalyptic Conspiracism

The interdependencies and shared attributes between apocalypticism/millennialism and conspiracism/conspiracy theories have been discussed for more than half a century in several academic disciplines and different contexts (e.g., Howard, 2011; Dittmer, 2010; Boyer, 1992; Barkun, 1986; Werly, 1977; Hofstadter, 1964). In his book *A Culture of Conspiracy*, first published in 2003, the American political scientist Michael Barkun (2013) describes "improvisational millennialism" as a practice in which individuals or groups draw from preexisting religious and non-religious beliefs to form new discourses in an improvisational manner which combine conspiracism and millennialism. Barkun (2013: 18) explains that this "third variety" of millennialism (in addition to religious millennialism or secular millennialism) is "distinctive for its independence from any single ideological tradition" (2013). Rather than relying only on religious texts or political beliefs or conspiracy theories, practitioners of improvisational millennialism draw from distinct ideological traditions, which can include, for instance, New Age, occultism, (fringe)science, conspiracism, religion, and radical politics. Here, newly formed conspiracy beliefs can even unify "domains which seem to be in opposition" (Barkun, 2013: 19), like American evangelical Christianity and New Age. The diversity of different traditions makes the arising millennial ideas attractive for a wide range of social groups because many individuals or groups can identify elements of their respective belief system in the newly constructed thoughts. At the same time, several events and developments can be explained by the new improvisational discourse due to the diversity of knowledge-claims. As the term suggests, improvising is a core practice of improvisational millennialism: whatever happens in the world can somehow be connected to the existing worldview. Wilson (2017: 423) describes improvisational End-Times discourses as a "developed form of syncretism [which] draws disparate elements together into a contingent but holistic ordering of meaning within an apocalyptic framework. It transcends and incorporates the discrete worlds of religious and secular apocalypses." Thus far, all of these characteristics also apply to apocalyptic conspiracism.

However, improvisational millennialist thought can be optimistic or fatalistic, or even both consecutively, when salvation is expected after disaster. As a specification or subset of improvisational millennialism, apocalyptic conspiracism only captures pessimistic and catastrophic conspiracist End-Time discourses, including those with an apparent religious origin and those without a clear religious emphasis. In this research, the emphasis is on the apocalyptic in a popular understanding, on doom and destruction, and not on the millennial hope related to a new beginning and a new earth. Barkun (2013: 225) writes concerning the relationship between millennialism and apocalypticism:

> Millennium and apocalypse tend to be used interchangeably, but in fact they possess different associations. Millennium connotes a time of maximal fulfillment—whether of prophecy, human potentiality, or divine promise. Apocalypse, according to Chip Berlet, suggests "an approaching confrontation, cataclysmic event, or transformation of epochal proportion, about which a select few have forewarning.

In this rather academic and religiously hopeful understanding of apocalypticism, the apocalypse is described as an event which initiates the millennium. Stewart and Harding (1999) write that the apocalypse is an abrupt and imminent crisis that ushers in a New World Order via a revelation or revolution manifest as the millennium. Traditionally, the term "millennium" meant the 1000-year period of Christian prosperity expressed in the book of Revelation. But more recently, the term "millennium" is rarely used in this "narrow and precise" sense (Cohn, 1970: 13). Rather, the millennium can also mean a radical transformation of revolutionary scale which destroys established power structures and establishes a new social order, "whether it is of a duration of forty years or four thousand" (Landes, 2008: 334). In short, millennium can mean an event, the transformation itself, but also the period of time that follows the transformation while the apocalypse can only mean the eventual lead-up and realization of this transformation.

Likewise, apocalypse has more meaning than simply the events which initiate the millennium. As mentioned in the outset of this introduction, the popular understanding of apocalypse describes the end-of-the-world scenarios, destruction, catastrophe, or disaster—not any change, but change *for the worse*. In the popular sphere, few would use the term "apocalypse" to describe a transformation into an age of prosperity. A zombie apocalypse unlikely refers to a transformative event after which humans and zombies enjoy life together (though at least two comedic films do suggest such a future, *Fido* [2006] and *Shaun of the Dead* [2004]). Leonardo DiCaprio's character in *Don't Look Up*, the scientist Dr. Randall Mindy, does not address any transformation of the world when he asks the American president: "You do understand that this is an apocalyptic event? This is a large celestial body headed towards our Earth." Rather, he means final disaster and the end of human life on Earth. And when Mike Hulme (2009: 345) addresses secular apocalyptic framings of climate change, he uses the term "apocalypse" "in its popular sense, meaning destruction." In the popular sphere, apocalyptic discourses do not mean just *any* transformation, but a transformation for the worse, destruction and the decay of contemporary societies. It is in particular this rather popular understanding of apocalypse which governs our conceptualization of apocalypticism, as the end of the current world precipitated through destruction and disaster (in addition to the meaning as revelation of knowledge described in the next section).

In short, this book discusses apocalypticism as "catastrophic millennialism" (Wessinger, 1997), the "turmoil" (Barkun, 2013: 255), chaos, and disaster, that precedes a *possible* millennium. When Robertson (2018: 236) uses the term "millennium" in millennial conspiracism to refer "to all accounts of more-or-less immanent planetary change, whether for better or for worse," we use the term "apocalyptic" in apocalyptic conspiracism to engage with planetary change "for worse" (2018), emphasizing the imminent resurgence of evil and prophecies of destruction and disaster where conspiratorially, to reiterate Barkun (2013: 255) above, only a "select few have forewarning." It is in particular the doomsaying and the pessimistic nature of the evangelical and non-religious texts of the American right that justify the use of the term "apocalyptic" in its popular understanding as disaster and destruction. Although the return of Christ and related events are sometimes addressed and framed as positive events within these evangelical digital spaces, the majority of the discursive content, including visual elements, emphasizes imminent doom and disaster. Nonetheless, as apocalyptic and millennialist belief systems are highly flexible and constantly updated to reflect contemporary events and political environments, the boundaries between millennial conspiracism and apocalyptic conspiracism can be rather fluid.

Conspiracist *Feindbilder*

Since conspiracy theories act as an "early warning system for group security" (Uscinski and Parent, 2014: 17) and "originate from feelings of existential threat" (van Prooijen, 2020: 23), the severe challenges imposed by climate change and the Covid pandemic contribute to the rising popularity of conspiracy beliefs. Explanatory models that assume malicious plots provide a narrative of overarching meaning and a form of security that helps to cope with uncertainties, anxious feelings, as well as the perceived loss of control caused by societal crises (Šrol, Ballová Mikušková, and Čavojová, 2021; Imhoff and Lamberty, 2020; Van Prooijen and Douglas, 2017). Yet, conspiracy discourses are more than just a coping strategy to deal with distressing change. Adherents of conspiracy theories hold that crises like climate change or Covid-19 are intentionally caused by malicious people or groups who profit from the crises themselves or from the connected political measures. Therefore, conspiracist beliefs provide an image of an enemy, someone to blame for the negative changes and disasters in one's life, it may be Bill Gates, the US Democratic Party, George Soros, the Illuminati, the Rothschild family, or Satan and the Antichrist.

In this book, we will employ the German term *Feindbild* (image of an enemy) to describe imaginations of the conspiring evil that are assigned to political, societal, or scientific actors and organizations. A *Feindbild* in conspiracy discourses emerges out of preexisting negative prejudices against social, religious, or ethnic groups, nation-states, or political ideologies. In times of crisis and existential threat, adherents of conspiracy beliefs draw from an established *Feindbild* of their social bonding group and cultural environment to project it onto individuals and institutions that are supposedly responsible for the conspiracy. Early academic engagements with the relationship between disaster, apocalypse, and conspiracy beliefs already highlight the relevance of imaginations of an evil, someone or something which can be blamed for the already experienced or prophesied catastrophe (Cohn, 1970; Hofstadter, 1964).

Conspiracist *Feindbilder* are also highly flexible. Established imaginations of evil can be embedded into different worldviews and projected onto a variety of individuals and organizations that fulfill the characteristics of the alleged evil.

For instance, conspiracy theories of the American right frequently portray the Democratic Party, including politicians like Hillary Clinton or Bernie Sanders, or environmentalists like Al Gore, as the evil conspirators that attempt to restrict the freedom of Americans, whereas the American Christian right projects their religiously founded *Feindbild* of the Antichrist or Satan onto the Democratic Party, liberals, and globalist institutions such as the United Nations. Adherents of millennial, apocalyptic, and conspiracy beliefs often divide the world in a dualistic way which only knows black and white, "us" and "them," good and evil, truth and lies. Barkun (2013, 1986) argues that this Manichaeanism embedded in conspiracist End-Times discourses emerges out of negative prejudices and acts as a populist simplifier of a seemingly increasingly complex world during times of crises. Moreover, such practices that project established images of an enemy onto contemporary relevant actors constitute a "demonization" of perceived hostile individuals, social groups, companies, or organizations (O'Donnell, 2020a, 2020b, 2019; Barkun, 1986).

The Apocalyptic Revelation of the Truth

One of the pivotal functions of apocalypticism is to provide exclusive access to the truth—meaning the "real" truth and not the specious "fake news" and "fake truths" disseminated by the mainstream media. The original Greek meaning of the term "apocalypse" (ἀποκάλυψις or *apokálypsis*), meaning "revelation or unveiling," hints at the epistemological dimension of apocalypse, which is to uncover hidden knowledge about the past, present, and future of humanity and Earth. In that context, DiTommaso (2020: 318) argues that "the primary operational function of apocalyptic speculation is to reveal the true nature of things to members of the group for which it is intended."

In that context, apocalypticists "believe to know" the truth about the past, present, and future of the world, as they consider themselves as a chosen group which possesses the key to the only true knowledge. Due to this understanding of apocalypticism, the concept lends itself well to conspiracism, as both discursive fields endeavor to unveil the actual truth about human history that is supposedly concealed by institutions in power. Those "in the know" therefore have access to truth and can therefore clear the fog for the unknowing mainstream and possibly guide a pathway to enlightenment and salvation (DiTommaso, 2020, Robertson, 2018, 2016; Megoran, 2012). Apocalyptic discourses separate societies into the knowing "elect" and the unknowing rest in a dualistic way. DiTommaso (2020: 318) explains: "One is either part of the Elect or is not: infidel or heretic. One cannot "agree to disagree" when eternal life (or equivalent) is at stake." This epistemological dualism also hints at the radicalization and political extremism of some apocalyptic groups: one is either with them or against them. Due to the revealing characteristics of apocalyptic conspiracism, it is important to conceptualize conspiracist End-Times discourses as knowledge or knowledge-entailing discourses, a politically relevant and powerful alternative imagination of the truth that increasingly enters the public sphere.

Conspiracist End-Times knowledge is not just any form of knowledge but in the light of established power-knowledge hierarchies of a society, it is a way of knowing that *counters* the dominant truths of a society. In this book, we mostly restrain from any judgments about what is true and what is false, but instead analyze how and why apocalyptic conspiracists construct their counter-knowledge and respective truths. It is not our intention to debunk any conspiracy beliefs, but we follow a social epistemological perspective that allows the existence of different truths that strive for power. We approach truth and knowledge as socially constructed, based on a theoretical perspective that assumes that any form of knowledge formation depends on subjective judgments and the social environment of the knowing individual. In this engagement with apocalyptic conspiracist ways of knowing about global crises, we argue that all individuals or social groups can only "believe to know," as any conception of objective truths is subject to power hierarchies, social negotiations, and subjective judgments.

However, a respective way of knowing about the world is also connected to moral judgments, political demands, and a strive for power. Here, a desired change of societal power hierarchies is always connected to establishing a certain way of knowing and its truths as the commonly accepted truth in a society (Foucault, 1980: 93). Hence, the dissemination of conspiracist End-Times thought by the American right must be understood as an attack on powerful political institutions, as their respective counter-truths imply certain political actions (e.g., if climate change is just a hoax, all climate protection policies and renewable energies can be abandoned). Ever since an odd coalition of right-wing conspiracists, evangelicals, esotericists, and anti-vaxxers hijacked the streets in North American and European countries to protest microchips in Covid vaccinations or Bill Gates' satanic population control measures, it became apparent that conspiracists are not an irrational minority which can be simply ignored but a vocal and powerful part of the political landscape. These conspiracists "believe to know" that Covid-19 and vaccinations are tools of population control, and endeavor to save themselves and the overall population from dangerous, possibly lethal, jabs. And ever since a QAnon shaman dressed in fur and bull horns stormed the American Capitol together with hundreds of MAGA supporters, fueled by Trump's conspiracy theory of a stolen election, most people became aware of the potential radicalism of conspiracism. Individuals who stormed the Capitol "believe to know" that Trump is the rightfully elected president and that Joe Biden and the Democratic Party plan the dissolution of the United States. Hence, QAnon supporters, apocalyptic evangelicals, and other conspiracists fight what they perceive as the evil: the Democratic Party, globalists, and the Deep State. These events show that an individual *Feindbild* (image of an enemy) is an important part of apocalyptic truths, often with dangerous implications for the individuals or groups who are believed to be the root of the evil.

Although many perceive the radical End-Times conspiracism of the American right as nonsense, given their conviction, such discourses are still a threat to democratic societies. Adherents of apocalyptic conspiracism believe that they are on the right side of history, and they believe they possess the knowledge to prove just that. Apocalyptic conspiracism is a way of knowing about the world that contradicts other ways of knowing about the world, typically the dominant or mainstream way of knowing about the world. But also, and crucially, the ways of knowing from Black, queer, and colonized subjects. Such apocalyptic conspiracism not only further marginalizes such groups but also targets such precarious minority groups "conjured through patterns of misogyny, homophobia, Islamophobia, antiblackness, and (settler) colonialism" (O'Donnell, 2021: 142). Contrasting perceptions of the truth are further connected to moral judgments about what is right and what is wrong. Adherents of pessimistic conspiracism perceive themselves as the good ones, the unsung heroes who sacrifice themselves for the greater good, but such salvation and dualistic thinking can have material a/effects on persecuted groups, and, as we explore later, climate action, change, and knowledge. However, in this book, and in no way adopting an apologist position, methodologically we explore apocalyptic conspiracist groups as part of the plurality of truth-claims and knowledge-producing systems which compete for power in society (Foucault, 1980). Moreover, we recognize that *Feindbild* does not do justice to the importance of these larger discursive constructs of "the foreign" and "evil" that maintain group identity, enable a positive self-image, and eventually constitute the foundation of apocalyptic beliefs.

Interventions

This book expands the literature on the discursive practices of anthropogenic climate change denialism and skepticism and further improves the understanding of religious, apocalyptic, and conspiracist belief systems, which affect the perception of geopolitics and global crises. In a 2024 survey, it was estimated that approximately 15 percent of Americans deny climate change, and the geography of that denial is clustered in areas with high white evangelical populations (Gounaridis and Newell, 2024), specifically the sunbelt of the United States (Dochuk, 2011). Only 31 percent of white evangelical Protestants believed climate change is caused by human activity. Donald Trump was the strongest influencer of climate denialist social media, and 59 percent of deniers polled during the 2024 primaries thought "that climate change is a conspiracy theory or hoax . . . and a shadowy attempt to dupe the public into bearing the costs of decarbonization" (Gounaridis and Newell, 2024: 5; cf. Hamilton, 2024). An emphasis on evangelical apocalypticism is of particular importance and urgency because the political power of conservative apocalyptic evangelicalism increased notably, in particular in relation to Trump's administration and the Republican Party more broadly (Gorski, 2019). Evangelical apocalyptic discourses can influence American political practices through networks of power and shape the geopolitical and environmentally relevant beliefs and behaviors of part of the American population (Sturm, 2010; Dittmer, 2010; Herman, 2001). In that context, our research illustrates how an anti-globalist apocalyptic geopolitical imaginaries fuel the rejection of global climate protection policies, while we conversely highlight how climate change becomes framed in a manner that reconfirms the trueness of preexisting American exceptionalist and anti-globalist beliefs. Through exploring climate change and Covid discourses of the religious and secular American conspiracist right, this book provides a comprehensive study of the field of American anti-globalism, highlighting the flexibility of apocalyptic conspiracist thought that can be adjusted to different crisis and to the beliefs of distinct cultural environments. Thereby, this research further contributes to the field of critical geopolitics as we show how geopolitical imaginations are discursively constructed and reinforced through knowledge-entailing discourses. Additionally, the examination of the field of apocalyptic conspiracism provides an insight into the epistemic strategies that sow powerful doubts about the legitimacy of science, the mainstream media, and official accounts of truth.

Theoretically, this book contributes to the academic literature in two ways. First, and as outlined before, we specify concepts of improvisational millennialism and millennial conspiracism to capture the pessimistic, fearmongering, and politically extremist apocalyptic conspiracist discourses that became increasingly popular and powerful in recent years. Second, we approach the internet as a network of digital spaces, as "discursive site[s] of practice" (Barr, 2011: 15), each with different prevailing truths, beliefs, ideologies, and epistemological preferences. Through transferring traditional theories of the geographies of knowledge onto the internet, we demonstrate that "space matters" (Livingstone, 2003: 5) not only in the construction, dissemination, and perception of knowledge in the "offline" world but also in the digital sphere, arguing for *digital geographies of knowledge*.

Overall, this research considers the interplay of different fields of public life that are relevant for the popularization and construction of apocalyptic conspiracist discourses. An important finding of our initial engagement with conspiracist thought is that authors use an assemblage of interconnected subjects to argue for the existence of an anti-Christian superconspiracy. A single piece of text, for instance, a blog article, can include references to the Bible, a critic of IPCC climate science, alternative empirical data that argues against human-made climate change, and a discussion of global politics, the Antichrist, and contemporary American politics. Since apocalyptic conspiracists construct a holistic argument, we hold that it is important to consider all parts of this argument. Although it is in particular the interaction of different subjects that provides power and legitimacy to apocalyptic conspiracist discourses, we split this book into three different parts that explore different aspects of the complex assemblage of apocalyptic conspiracism and discuss examples in the light of relevant literature.

The first part of this book provides the theoretical foundation and methodology for the analysis of apocalyptic conspiracist discourses. Chapter 2 introduces the academic field of social epistemology as well as a theoretical perspective that conceptualizes all truth-claims and all forms of knowledge as social institutions. This book approaches truth and knowledge as socially accepted beliefs (Shapin, 1994) and further argues that imaginations of the truth and valid scientific practices differ across cultural environments, spaces, and societies (Livingstone, 2002a; Ophir and Shapin, 1991). After introducing key concepts of the field of social epistemology and geographies of knowledge, the second chapter further highlights a plurality of truth-claims, knowledge, and knowledge-producing systems which compete for power in society (Foucault, 1980). The resulting power-knowledge hierarchies are highly relevant for the generation of End-Times thought since knowledge and knowledge-creating practices that are suppressed or dismissed by the institutions in power constitute the foundation of apocalyptic conspiracism. Following Barkun (2013) and Robertson (2018, 2016), the second chapter emphasizes that *apocalyptic conspiracist counter-knowledge* arises out of preexisting beliefs and established ways of knowledge creation of the particular social environment out of which counter-knowledge arises. To provide an example of the knowledge that is "countered," Chapter 2 provides a critical discussion of the IPCC climate change knowledge as this book addresses all forms of knowledge, including this dominant or mainstream anthropogenic climate change knowledge, as shaped by society, space, and social power relations (Mahony and Hulme, 2018; Demeritt, 2001).

Based on the assumption that different spaces enable distinct perceptions of the truth, Chapter 3 explains that the internet and modern communication technologies enable digital spaces (Ash, Kitchin, and Leszczynski, 2018; Graham, 2014), in which knowledge-claims can be created, discussed, disseminated, and consumed. Since the particular cultural, social, and technological characteristics of a digital space affect the generation of knowledge and truth-claims, the third chapter argues for the need to scrutinize *digital geographies of knowledge* as a way to analyze the respective truth-claims and truth-generating practices of online communities and social bonding groups that exist in digital spaces. Due to the relevance of the internet in propagating conspiracist End-Times thought (Barkun, 2013), the construction of counter-knowledge discourses is examined by analyzing relevant primary sources that emerge from digital apocalyptic conspiracist spaces. The collection of primary data from a range of digital sources such as alternative news websites, evangelical apocalyptic blogs, videos, and podcasts is described in the third chapter.

In the next three chapters, we explore American evangelical apocalypticism and the construction of evangelical conspiracist climate change discourses. Before engaging with the analysis of the collected primary data, Chapter 4 introduces the diversity of American evangelicalism, its conservative and Christian nationalist movements, as well as the power of evangelicalism in the American society. We investigate American evangelicalism not just as a religious movement but also as a political and cultural movement, as this book rejects a clear distinction between religion and politics (Asad, 1993; Fitzgerald, 2011). Based on a Bourdieuan approach to religion that assumes a reciprocal relation between religious and non-religious dimensions of society, we argue that all forms of evangelical knowledge, including knowledge concerning the climate and natural environment, emerge out of an assemblage of religious, political, cultural, and scientific dispositions. After introducing traditional concepts of different millennial belief systems of American evangelicalism and their impact on environmentally relevant beliefs, we focus on dispensational premillennialism, an apocalyptic belief system that constitutes one religious foundation of what we call apocalyptic conspiracist discourses. In that context, the fourth chapter further defines the key beliefs and practices of American evangelical apocalyptic conspiracism.

Chapter 5 provides an analysis of the social, cultural and political arguments of evangelical apocalyptic conspiracist climate change discourses and its relation to non-conspiracist and non-apocalyptic forms of evangelical climate change skepticism and denialism. We present primary data that illustrates that preexisting premillennial dispositions influence the perception of climate change and related discourses, whereas evangelical counter-knowledge on climate change and environmental politics reconfirms the trueness of the holistic religious conspiracist belief system. In that context, we also discuss how apocalyptic conspiracist authors frame climate change in a manner that reinforces preexisting *Feindbilder* of apocalyptic evangelicalism. Relating to the relevance of power-knowledge hierarchies discussed in the second chapter, Chapter 5 further explains how evangelical conspiracists portray themselves as marginalized by institutions in power as they somewhat use the tools of social constructionism to engage with the discursive practices of the perceived climate change propaganda machine.

Following the emphasis on the sociocultural dimension of conspiracist End-Times discourses in the fifth chapter, Chapter 6 explores the counter-hegemonic knowledge-creating strategies through which authors attempt to disprove the existence of anthropogenic climate change. Since not only truth or knowledge-claims but also knowledge-creating strategies are inscribed in space, the sixth chapter of this book presents the epistemic practices that are portrayed as valid and legitimate ways of creating knowledge in evangelical apocalyptic conspiracist digital spaces. We explore how authors refer to the Bible to argue that humans do not have the power to alter or destroy God's creation. Based on the widespread evangelical assumption that the Bible and science produce the same truths, anthropogenic climate change denialist authors additionally refer to a wide range of scientific climate change denialist sources that are believed to provide empirical evidence for the non-issue of climate change. Another knowledge-creating practice addressed in Chapter 6 is the use of anecdotal evidence and personal first-hand experiences.

The focus of this final chapter concerns the geopolitical beliefs embedded in conspiracist End-Times discourses. We argue that the imagination of an evil one-world government drives much of the resistance to restrictive crisis politics. Chapter 7 explains how evangelical authors utilize international climate politics to reinforce preexisting anti-globalist beliefs of evangelical apocalypticism. In that context, we discuss, for instance, how Trump's withdrawal from the Paris Climate Agreement was employed to strengthen religiously motivated American exceptionalist beliefs. We also illustrate how the climate-skeptic attitudes of the Trump administration are portrayed as proof for the uniqueness of the United States, as God's chosen nation that has the power to resist the anti-Christian climate change deceptions. We conclude that climate change discourses reinforce a Manichean spatial imagination in which the United States, as the good side in the apocalyptic conflict, must fight against the rest of the world and American globalists who have been lost to the pressures of the Antichrist.

We conclude our argument laid out in this introduction by elaborating on the similarities between the apocalyptic conspiracism of the American right and environmentalist eco-apocalypticism, which are usually thought to exist in different political and cultural landscapes. Our final remarks put the effects of apocalyptic conspiracism into perspective through the 2024 Trump presidential campaign. In early 2024, Trump gave his listeners a Manichean ultimatum: "You can be loyal to American labor, or you can be loyal to the environmental lunatics." Continuing his America First narrative of "real" Americans versus liberals and globalists, Trump enforces a narrative in which you can only support one side, where his electorate or target group will likely identify with the American workers, thereby demonizing the environmentalists. Ringing of the geopolitics of his Republican predecessor, President Bush, one is either with us (America) or against us (the rest of the world). We conclude that apocalyptic conspiracism of climate change is a powerful social epistemology in US politics. In tracing how apocalyptic conspiracism is networked through religion and secular digital spaces, we therefore conclude—resisting prophetic temptation—with a call to further develop how such discourses function and are instrumental in the halls of power if we want to prevent such counter-epistemic discourse from becoming the dominant episteme.

2

Social Epistemology, Power, and Climatic Counter-Knowledge

A term that is often used in the context of right-wing populism, conspiracism, and alternative ways of knowing, is "post-truth." Through Donald Trump's 2016 presidential campaign and his use of conspiracy beliefs, "the terms 'post-truth' and 'post-fact,' virtually unknown 5 years ago, have exploded onto the media scene with thousands of recent mentions" (Lewandowsky, Ecker, and Cook, 2017: 354). *Oxford Dictionary* designated "post-truth" as the word of the year 2016 and defines it as "relating to or denoting circumstances in which objective facts are less influential in shaping public opinion than appeals to emotion and personal belief" (Oxford Dictionary, 2016).

This definition, however, presupposes that there are "objective facts" that can expose the lies of the post-truth world. Yet, it is important to consider that *all* forms of facts arise "out of particular configurations of practices, discourses, epistemic politics and institutions" (Sismondo, 2017: 3). This also applies to the "objective fact" in the Oxford Dictionary's definition of post-truth, which, according to a rather social constructionist perspective, is hard or even impossible to acquire, as all forms of knowledge, including scientific facts, are affected by preexisting beliefs, cultural histories, and the social environment out of which knowledge arises (Harambam, 2021; Latour, 2004; Shapin, 1994; Latour and Woolgar, 1979). Following such a theoretical perspective, the Oxford Dictionary's definition of "post-truth" is insufficient in the sense that it assumes that a fully objective truth can be agreed upon. Rather, the definition of an objective truth is always a subjective judgment. A society produces a hierarchy of different facts and knowledge-claims as different ways of knowing the competition for power in a society. What constitutes a purported objective truth and what is just a lie or a conspiracy theory is not just any subjective judgment but a judgment by institutions with a high epistemic authority, like universities, or in the case of climate change, mainly the UN's IPCC.

Based on this social constructionist perspective, our research does not debunk any apocalyptic conspiracist discourses, as the practice of debunking requires that we are in possession of the objective truth. Discussing if academics should debunk conspiracy theories, Harambam (2021: 105) following Max Weber's essays on sociology, states that "debunking conspiracy theories is not possible, not professional and not productive." A highly relevant reason here is that any form of evidence that would disprove a conspiracist knowledge-claim is considered to be part of the conspiracy. But more important is the impossibility of an unbiased judgment on the truthfulness of conspiracy narratives. Accusing conspiracists of disseminating lies is a subjective judgment, a practice that does not defend the scientific neutrality and impartial research but a practice that contradicts any neutrality. Therefore, "if scholars are to maintain the appropriate type of neutrality, they should not set out to debunk conspiracy theories" (Hagen, 2020: 435). Just like conspiracist knowledge, our knowledge, and any knowledge of our world, "is the product of our own meaning-making practices" and hence, "nobody can claim to know the real, objective, and only truth about the world we live in" (Harambam, 2021: 109).

Instead of debunking conspiracy discourses, our research asks *how* and *why* certain people and social groups reject dominant ways of knowing and create knowledge-claims that counter the supposed "objective truths." Following Harambam's (2021: 109) approach to conspiracy discourses, we ask, "How and why different people in different cultural contexts create and lend authority to (their versions of the) world?" To do so, we employ a social epistemological theoretical framework. The first section of this chapter presents key definitions and concepts of the field of social epistemology. Following Foucault (1980) and Robertson (2016), the emphasis of this section is on the plurality of truth-claims and the resulting power-knowledge hierarchies. Based on the concept of societal power-knowledge hierarchies, we conceptualize apocalyptic conspiracist discourses as subjugated or suppressed knowledge that counters the dominant or powerful knowledge in a society (Barkun, 2013). As this book argues that all forms of knowledge are the result of cultural histories and social power relations, the second section critically engages with IPCC climate knowledge and the generation of the supposed "objective facts" concerning climate change. Therefore, we also define the dominant and mainstream knowledge concerning climate change that gets countered by apocalyptic conspiracists.

Section I: Social Epistemology

All major areas of research of this book—apocalypse, conspiracy, geopolitics, as well as the science of climate change and Covid—can be approached as forms of knowledge, generated and communicated through discourse. Apocalyptic discourses are a "mode of knowing" (Stewart and Harding, 1999: 286) that reveals truths on future change as well as covert past and present-day processes that prepare the allegedly imminent apocalyptic transformation (DiTommaso, 2020; Williams, 2010; O'Leary, 1998). Conspiracist discourses are more than just an "alternative reality" (Bjerg and Presskorn-Thygesen, 2017: 138) but also an increasingly powerful form of knowledge that influences public political debates and the way people perceive the world around them (Robertson, 2018, 2016; Barkun, 2013; Jones, 2012). This also applies to geopolitical conspiracy discourses as they are conceptualized as a "body of knowledge" (Jones, 2012: 53) that claims to provide the reality of large-scale political geographies (Johnson-Schlee, 2019). Geopolitics in general can be approached as "constellations of knowledge" (Ó Tuathail, 1998: 16) or "abstract forms of knowledge" (Müller, 2008: 329) that disseminate truths about international politics and power relations of global space. Hence, a suitable theoretical framework to engage with scientific and geopolitical apocalyptic conspiracist truth-claims is one that engages with the origin, formation, dissemination, and justification of knowledge and its power in society: a social study of knowledge, which is most commonly described as *social epistemology*.

Epistemology (episteme/ἐπιστήμη = Greek for knowledge) can most broadly be defined as "an account of knowledge" (Moser, 2002: 3), "theory of knowledge" (Goldman, 2002: 164), or simply, how we know what we know. Epistemological studies investigate *how* we know and examine the nature, origins, and validation of knowledge, as well as its justification (Moser, 2002). *Social epistemology* emphasizes the social dimensions of knowledge-making processes as well as knowledge dissemination (Goldman, 2002). In social epistemology, epistemic subjects are treated as "socially constituted individuals who stand in relations of power" (Fricker, 1998: 134), although social power enters epistemic practices to varying extents. Historical, social, and geographical considerations relating to the plurality of knowledges and knowledge hierarchies (Fricker, 2010, 2007, 1998; Burke, 2000; Foucault, 1980), testimony (Goldman, 2002; Craig, 1990), and trust (Jasanoff, 2010; Shapin, 1994) are relevant themes within the social studies of knowledge. Thereby, social epistemology challenges rather traditionalist perspectives within epistemology which see science, knowing subjects, and knowledge-producing subjects as disconnected from political and social power relations.

Knowledge and Truth as Social Institutions

The general definitions of epistemology and social epistemology rely on the term "knowledge," a term which is interpreted in several ways depending on discipline and theoretical tradition. Addressing debates about varying understandings of knowledge, the American philosopher Alvin Goldman (2002: 183) differentiates between "strict senses of knowledge" and "loose senses of knowledge." The main characteristic which distinguishes these two understandings of knowledge is the assumption of an objective and absolute truth. Strict senses of knowledge require a belief to be absolutely true, or justified and absolutely true, to count as knowledge, whereas loose senses of knowledge describe commonly accepted beliefs as knowledge.

The conceptualization of knowledge in a loose sense is largely guided by a social constructionist perspective on knowledge. Loose senses of knowledge are often used within the history and sociology of science, for instance in the work of Steven Shapin. In his historical studies of scientific practices, Shapin explains that knowledge is inextricable from social relations (Shapin, 1994; Shapin and Schaffer, 1985). What counts for a community as valid knowledge, Shapin (1994) argues, is a collective good and collective accomplishment produced by the truth-generating practices of social groups. In relation to the term "truth," it must be noted that Shapin (ibid: 6) defines truth as "a social institution" which is generated out of commonly "accepted belief[s]" (ibid: 4) of social groups. Hence, knowledge in a loose sense cannot be described as entailing an objective and absolute truth since truth itself is conceptualized as a social construct. Overall, Shapin (ibid: 7) highlights the "collective nature" of knowledge in which common and established beliefs constitute what is considered as valid and legitimate knowledge.

The collective nature of knowledge is also closely tied to space as the truth-generating practices of social groups are locally shaped (Ophir and Shapin, 1991). Hence, space is epistemologically significant. In relation to this, Livingstone (2002a: 12) argues that science, as a knowledge-making practice, needs to be understood as a "cultural formation with local practices that are rooted in a range of particular spaces." Geography influences knowledge-making in a variety of scales, from small-scale spaces of scientific enquiry like a laboratory and their respective spatial particularities toward the scale of entire nations and cultures (Jasanoff, 2010; Livingstone, 2003, 2002a). The multiple relationships between space, place, social and cultural environments, and knowledge are established areas of inquiry in science and technology studies as well as in the academic field of the *geographies of knowledge* (Gregory, Meusburger, and Suarsana, 2015).

Shapin's thought is generic for a revisionist stream within social epistemology and social studies of science which questions the objectivity of truth (Goldman, 2010) by arguing that knowledge-claims, facts, and truths arise out of social, cultural, historical and geographical configurations (Shapin, 1994; Ophir and Shapin, 1991; Shapin and Schaffer, 1985). Goldman (2002: 185) termed Shapin's understanding of knowledge as socially accepted belief as "institutionalised knowledge" or a social institution. While Goldman (1999, 2002) argues to employ a strict sense of knowledge and thereby follows the most realist philosophical epistemologists, he also acknowledges that specific research questions, case studies, and academic disciplines appropriately employ a rather social constructionist conceptualization of truth and knowledge. Following Barkun (2013), Robertson (2018, 2016), Hagen (2020), and Harambam (2021), we hold that the engagement with conspiracy discourses is one of the research fields in which it is justified to employ a theoretical perspective which rejects strict senses of knowledge and conceptions of "objective truths" but instead approaches all forms of knowledge in a loose sense in which individuals and social groups "believe to know."

Plurality of Knowledges, Epistemes, and Knowledge Hierarchies

The main concern of this research is an analysis of white American apocalyptic conspiracist knowledge-claims which contradict the dominant perceptions of the truth concerning geopolitics and climate change. In that context, this research requires not only the notion of a plurality of truths and knowledge-claims but also of a plurality of knowledge-producing systems out of which contradicting knowledge-claims emerge. Much of the academic work we use to approach contrasting ways of knowing about the world is based on Foucault's concept of an *episteme* and other works by the French philosopher, which argue that any form of truth, knowledge, or fact is to be understood as the result of social processes and power relations.

In Foucault's (1980: 197) later writings, an episteme is described as a "discursive apparatus" which does not determine what is considered to be true or false but which regulates the scientificity and legitimacy of knowledge- and truth-claims (Mahony and Hulme, 2018; Robertson, 2016; Foucault, 1980; Foucault, 1990). An episteme influences knowledge-making practices (*epistemic practices*) and dispositions that affect the perception of knowledge, as it is "governing the way in which people understand, and act in, the world" (Bevir, 1999: 347). Epistemes are not knowledge or truth-claims themselves, but rather the established practices which govern the production and perception of truth and knowledge. Foucault originally argued that each epoch had its own episteme (see Foucault's *The Order of Things,* 1970). Yet, Foucault's (1980) later notion of epistemes describes an episteme as a knowledge-producing system composed out of respective sets of rules which can coexist simultaneously in a society without clear temporal distinctions. What is important for the analysis of apocalyptic conspiracist knowledge is that epistemes and their respective perceptions of the truth compete against each other as they strive for political power. Here, powerful social groups potentially denounce or suppress certain knowledge-claims or knowledge-creating strategies (Fricker, 2007; Burke, 2000). This results in an epistemic power asymmetry or a *hierarchy of knowledge* (Foucault, 1980). In that context, Foucault (ibid: 131) explains that "each society has its regime of truth." A *regime of truth* is a rather abstract construct which significantly contributes to a hierarchy of knowledges within a society. It is, again, to be understood as a discursive "apparatus" (ibid: 132) which is constituted out of the discourses which are accepted as true by a society. While an episteme is an "apparatus which makes possible the separation . . . of what may from what may not be characterised as scientific" (ibid: 197) by establishing knowledge-producing strategies, the regime of truth manifests societal mechanisms of power which distinguish what is believed to be true from what is believed to be false.

However, any episteme, regime of truth, and their power relations are "inscribed in place" (Ophir and Shapin, 1991: 12; following Foucault, 1980) and consequently deviate from each other subject to different spatial conditions. In general, Foucault (1980) draws attention not only to historical but also to spatial disparities of knowledge-power systems since the discursive apparatuses which characterize an episteme or a regime of truth vary depending on geography. Across different communities, social groups, or nations, there are a number of "apparatuses for making and contesting truth" (Jasanoff and Simmet, 2017: 753) that compete for power and govern the validity and legitimacy of knowledge and truth-claims.

Foucault (1980) further highlights an imbalance between the consumers and producers of the truth as he explains that true knowledge is consumed and diffused by society as a whole but produced dominantly by a small number of powerful apparatuses. These powerful apparatuses operate through "epistemic authorities" (Robertson, 2016: 46), such as universities or governmental institutions. Epistemic authorities relate to the regime of truth as they have the power not only to produce knowledge but also establish what is to be considered as valid, true knowledge through discursive practices. The power imbalance between producers and consumers of knowledge is also critically addressed in Harambam's (2021: 116) engagement with conspiracy discourses. Harambam argues that a democratization of knowledge and an increased epistemic diversity, for instance, through "deliberative citizen knowledge platforms" that assess the quality of public information, could decrease the contemporary conspiracist skepticism. Criticizing the power of epistemic authorities, Harambam (2021: 117) further writes that "our democratic societies, and the knowledge and values we want to live by, are too important to be left in the hands of a powerful elite of experts and technocrats alone."

However, the use of language and other discursive practices are among the "subtle mechanisms" (Foucault, 1980: 102) of power through which societies organize knowledge- and truth-claims and constantly reproduce a hierarchy of knowledge. One of those subtle mechanisms is addressed by Shapin (1994), who argues that, as a discursive practice, the mere use of terms like "knowledge," "truth," or "facts" entail a judgmental character which represents a sorting mechanism, contributing to the establishment of knowledge hierarchies. Beliefs which are commonly accepted are described as "knowledge," "facts," or "truth," whereas beliefs and statements not accepted by a community are sorted out or downgraded by using a different terminology.

Especially in current Western political discourse, the work of Foucault (1980) and Shapin (1994) is highly relevant as powerful political actors attempt to denounce knowledge-claims by political opponents through language and discourse (e.g., "fake news," "lies") to establish a hierarchy of knowledge in their favor. This example also illustrates that "truth is linked in a circular relation with systems of power" (Foucault, 1980: 133)—a reciprocal relationship between power and truth in which powerful subjects of a society can establish what is to be seen as truth through discourse while conversely benefiting from this truth to sustain or increase their power. Knowledges at the bottom ends of knowledge hierarchies, typically produced by less powerful groups and their respective epistemic strategies, are termed "subjugated knowledges" by Foucault (1980) or "suppressed" and "stigmatized" knowledges by Barkun (2013) in improvisational millennial or, in our conceptualization, apocalyptic conspiracist contexts.

Justified Knowledge, Trust, and Epistemic Capital

Social epistemology engages not only with the generation of knowledge and power-knowledge hierarchies but also with the perception and justification of externally acquired knowledge. Most beliefs and perceptions of the truth rely on information which is "external to the believer's cognitive perspective" (BonJour, 2002: 234). This is, for instance, evident in the case of climate change. The complexity and interconnectivity of the earth's spheres, the technical requirements to collect relevant data, or the computer power to process large amounts of climate data, make it difficult for laypeople to engage with scientific practices of climate science or generate their own reliable global climate change knowledge. To put it succinctly, climate is "everywhere and nowhere" (Jasanoff, 2010: 237). For instance, greenhouse gases are "beyond human senses" (Dalby, 2013: 43) and not detectable by humans without appropriate technical equipment in a sense which allows scientific research. Hence, individuals, political decision-makers, and society as a whole are "epistemically dependent" (Fuller, 2002: 278) on experts and large knowledge-making institutions, such as the IPCC, who accomplish creating meaningful knowledge on the climate through a synthesis of a vast amount of research conducted on the earth's climate, ecology, and connected systems (John, 2017).

An internalist stream within epistemology would argue that the majority of climate change knowledges laypeople hold are not justified. *Internalism* holds that knowledges are only justified if "all of the elements needed for a belief" (BonJour, 2002: 234) are acquired through first-hand, cognitive experiences. However, as outlined before, it is virtually impossible for an individual subject to acquire reliable and robust knowledge on climate change (and other highly complex scientific issues, e.g., the developments of mRNA vaccines) through first-hand, cognitive experiences. Therefore, we will follow Shapin (1994: 9), who argues that acquired knowledges can be considered legitimate and justified if an individual trusts an external source of knowledge to be a "reputable and veracious" source of knowledge. Shapin (1994: 7) defines *trust* as the "moral bond between the individual and other members of the community." Trust, for Shapin, is always present in knowledge production and consumption since individuals, social groups, or societies must trust previously created knowledge to attain new knowledge. Without a trustful acknowledgment of previous knowledges, knowledge-producing individuals would have "to work backwards through their community's accumulated knowledge" (1994: 19) to validate all prior knowledge-claims. In the absence of trust, there would be a "void" (1994: 27) without any prior knowledge to build on. Shapin (1994: 16) then disagrees with internalist epistemologists, who "systematically argued that legitimate knowledge is defined precisely by its rejection of trust."

Trust also largely influences what is perceived as an epistemic authority and it governs which experts to believe and which ones to reject (Fricker, 1998, following Craig, 1990 and Shapin, 1994). In that context, the judgment on the trustworthiness of experts is influenced by a "range of human values, including political and religious ones" (Suhay and Druckman, 2015: 7). Goldman (2002: 173) calls these trust-influencing values "acceptance dispositions," which are formed by the particular cultural and social conditions of a community (Goldman, 2002; Fricker, 1998). Acceptance dispositions increase the likelihood for accepting knowledge, but they can also result in the rejection of knowledge- or truth-claims due to the absence of trust or distrust. A lack of trust during the consumption of knowledge can be caused by an identity prejudice toward creators and communicators of knowledges.

Fricker (2007: 18) explains that consumers of knowledges "must make some attribution of credibility regarding the speaker" and, moreover, that this credibility attribution is influenced by social identity. Here, an identity prejudice can cause a consumer of knowledge to reject it due to the communicator's social identity, characterized for instance, by gender, race, political affiliation, religion, or class. Identity prejudices result in lower levels of trustworthiness and credibility. Still, identity prejudices "can come in positive and negative form[s]" (2007: 27–8). In a positive identity prejudice, consumers of knowledge are more likely to trust members of their social bonding group (Goldman, 2002). Religion, for instance, might work as a positive identity prejudice since religious people are more likely to accept knowledge-claims produced by members of the same religious community, especially when knowledges are based on established epistemic strategies of that community. Nonetheless, the acceptance of testimony is not restricted to speakers with close social ties to the hearer. Generally, trust, and thereby the acceptance of experts, is largely influenced by group identity, a point which Shapin (1994) also highlights. Here, geography influences who is perceived as a trustworthy communicator of knowledge. Shapin (1998: 8) further argues that trust depends on geography: "Whom to trust? Answers to such a question will vary from place to place and from culture to culture."

In the context of climate change, acceptance dispositions and identity prejudices can cause individuals or communities to reject the global knowledge of anthropogenic climate change as it was produced by creators, which they generally distrust and, hence, rather turn to knowledges created by their social bonding group. We will show that acceptance dispositions, perceptions of experts, and trust are of particular relevance for American apocalyptic conspiracist communities when deciding on which knowledge to accept and which to reject.

Trust, acceptance dispositions, and identity prejudices also influence what we call epistemic capital, following Bourdieu's (1998) comprehension of capital and Robertson's (2016) work on the epistemology of millennial conspiracism. In Bourdieuan theory, different individuals or groups within a societal field compete over the production, possession, and consumption of a variety of forms of capital. Bourdieu was influenced by Marx's notion of economic capital but added other forms of capital such as social capital (e.g., a network of relationships which can be used to increase one's power in society) and cultural capital (e.g., relevant knowledge or skills).

Robertson (2016) expanded Bourdieu's idea of capital by adding epistemic capital. Epistemic capital "does not map what you know, but how you know" (2016: 29). Here, we want to emphasize the power of epistemic capital, where, consistent with Bourdieu, "capital is equivalent to power" (Rey, 2014: 52). We define epistemic capital as the power of knowledge: the power of establishing societal perceptions of the truth and the power to influence politics and social and cultural practices through the validation of certain knowledge-claims and knowledge-creating practices. The power to govern which epistemic strategies, perceptions of the truth, and knowledge-claims are considered as valid and legitimate depends on the epistemic capital of an institution, individual, or social group. But just like other forms of capital, there is an uneven distribution of epistemic capital within a society. The above-mentioned imbalance between the consumers and producers of the truth can be addressed by the concept of epistemic capital, as only a few individuals or institutions possess enough epistemic capital to affect societal debates and political decisions. In that context, conspiracism needs to be understood not only as a practice that challenges epistemic authorities and political institutions but also as a practice that aims toward claiming (or accumulating) epistemic capital and the means of knowledge production.

In this context, Verter (2003) illustrates that the power of Bourdieu's capital changes depending on the respective societal field, social environment, and space. An epistemic authority and its respective epistemic strategies and truth-claims might be powerful in one space, like a certain nation, but relatively powerless in another one because the preexisting acceptance dispositions of this cultural space result in a lack of trust toward this epistemic institution. For instance, the IPCC's epistemic capital might be low in evangelical apocalyptic conspiracist communities which reject anthropogenic climate change knowledge because of missing trust and negative identity prejudices toward intergovernmental (scientific) institutions. Still, the IPCC has the power to influence climate politics on a global scale, which shows that epistemic capital depends not only on space but also on scale.

Jasanoff (2010: 239) explains that different "civic epistemologies" as "state-centred methods of validating knowledge" (2010) tended to exist. Here, trust, acceptance dispositions, and the general creation and consumption of knowledge could be analyzed on the scale of nation-states. But today, civic epistemologies are challenged by new forms of social organization in which "earlier settlements between natural knowledge and political order" are dissolved. For instance, the diversity of political, cultural, and religious dispositions and the resulting contrasting imaginations concerning the truth of climate change and Covid in the United States illustrate that there is no civic epistemology, no unifying way of knowing about the world in the United States. The epistemic diversity within the same geographical region requires the conceptualization of "abstract spaces" (Livingstone, 2003: 6) as "intellectual, social, and cultural arenas" (2003) to analyze differing acceptance dispositions, trust, epistemic capital, and other factors which govern the production and consumption of knowledge, in particular in the context of an increasingly digitalized world (see Chapter 3).

Apocalyptic Conspiracist Counter-Knowledge

In light of the above-mentioned power-knowledge hierarchies, Barkun (2013) conceptualized denounced counter-knowledge as stigmatized knowledge. *Stigmatized knowledge,* or a *stigmatized knowledge-claim,* is an important source for improvisational millennialism and, thereby, apocalyptic conspiracist discourses. The concept blends in well with the social epistemological perspective of this book, which discusses valid knowledge, or the truth, as socially constructed and tied to social power relations. Stigmatized knowledge-claims are "claims to truth that the claimants regard as verified despite the marginalization of those claims by the institutions that conventionally distinguish between knowledge and error—universities, communities of scientific researchers, and the like" (Barkun, 2013: 26). In other words, stigmatized knowledge-claims are knowledge-claims which are regarded as true by some individuals or in some cultural milieus but denounced as false by a society on a large-scale, epistemic authorities, and political institutions in power. Here, geography is relevant again, as in some "cultural arenas," something is considered the truth while it is considered a lie in other spaces. Stigmatized knowledge is not objectively invalid or false knowledge by definition, but it becomes stigmatized by powerful epistemic institutions through discursive practices (Robertson, 2016; Foucault, 1980). The stigmatization is a result of power-knowledge hierarchies since epistemic authorities have the power to denounce certain knowledge-claims while conversely, the stigmatization of concurring knowledge-claims and epistemes reinforces the dominant episteme and its institutions.

Epistemic authorities do not only make judgments on the validity of certain knowledge-claims, but they also govern which knowledge-creating strategies (epistemic strategies) can produce warranted knowledge. Robertson (2018, 2016) uses Barkun's work on improvisational millennialism as well as theories by Bourdieu and Foucault to define knowledge-creating strategies which are rejected by epistemic authorities as *counter-epistemic strategies* or *counter-hegemonic epistemic strategies.* Robinson's (2016) idea of counter-epistemic strategies is suitable to examine apocalyptic conspiracism because the conspiracist discourses of interest for this book do not attack a single knowledge-claim but an entire episteme and its political and knowledge-creating institutions. It is usually not argued that just the science of anthropogenic climate change is false, but rather that the entire apparatus of mainstream science deliberately produces false knowledge to reinforce or increase the power of the assumed conspirators. For instance, epidemiologists and epistemic authorities like Anthony Fauci are attacked as conspirators who deliberately disseminate false information (see Part II of this book).

Counter-epistemic strategies can be derived from traditional, indigenous, or apocalyptic cultural milieus and are often located at the bottom end of a society's knowledge hierarchy. Counter-epistemic knowledge often entails counter-elite traits, which Foucault (1980: 84) describes as an "insurrection of knowledges" which does not only oppose the dominant knowledges or dominant epistemic strategies but also the "centralising powers which are linked to the institutions and functioning of an organised scientific discourse" (1980). This counter-elitism is also found in apocalypticism as a form of resistance literature that promotes the idea that only select few have access to the actual truth about the past, present, and future (DiTommaso, 2020; Megoran, 2012). Driven by their exclusive knowledge about the truth, apocalyptic conspiracists "believe to know" that the majority of society must be wrong about important issues like climate change and Covid because it is a fundamental creed of apocalypticism that only "the Elect" can be in possession of the truth due to their alternative, or counter-epistemic, ways of generating knowledge.

Nevertheless, conspiracy arguments (apocalyptic or not) are increasingly propagated by powerful political elites, as most prominently illustrated by the "presidential conspiracism" of ex-president Donald Trump (Muirhead and Rosenblum, 2019). As the so-called "denier in chief," the former American president Donald Trump frequently questioned the existence of anthropogenic climate change, stating that the science is faulty or that cold and snowy days disprove the existence of global warming (De Pryck and Gemenne, 2017). But in addition to these climate change denialist arguments, he also refers to conspiracist arguments, tweeting in 2012 that the concept of climate change was only "created by the Chinese in order to make U.S. manufacturing non-competitive" (Trump, 2012 in De Pryck and Gemenne, 2017). Likewise, other Republican elites, like late-senator James Inhofe from Oklahoma, called climate change *The Greatest Hoax*, the title of his 2012 book, which some have interpreted as an instrumental ideology to justify conservative political ideologies that downplay the threat of global warming (Uscinski, Douglas, and Lewandowsky, 2017). The most prominent example of presidential conspiracism might be Trump's claim that the 2020 presidential election was rigged to benefit Joe Biden and the Democratic Party, a claim that was also embedded into the millennial and apocalyptic conspiracist belief systems of QAnon and the convergent contingency among the evangelical right.

Yet, these conspiracist knowledge-claims by political elites, like the American president, still fall into the realm of counter-knowledge. Regardless of who is actually in position of power, conspiracists portray their respective knowledge as suppressed or subjugated by evil elites. Populist politicians like Donald Trump employ conspiracist discourses not only simply to demonize political opponents, often based on an established *Feindbild* of the potential electorate, but also to create a sense of crisis (regardless of its actual existence), purportedly caused by "political elites," to rally for support and stage themselves as the solvers of the supposed crisis (Pirro and Taggart, 2022). Even if conspiracists like Trump are in power, they still construct themselves as suppressed by a liberal mainstream and/or the globalist Deep State that allegedly controls American politics and the media.

Section II: The IPCC as the Dominant Climate Change Knowledge

The conceptualization of apocalyptic conspiracist discourses as counter-knowledge requires the designation of the dominant knowledge that gets countered. Power-knowledge hierarchies are constantly restructured and reinforced, and what constitutes dominant knowledge, and counter-knowledge, always depends on time, scale, and space. Yet, in the case of climate change, there is a form of historically established, powerful global knowledge disseminated by an institution with a high amount of epistemic capital that shapes dominant truths on climate and environmental change: the IPCC (Hughes, 2015).

The IPCC was established in 1988 by the United Nations Environment Programme (UNEP) and the World Meteorological Organization (WMO), first, to assess scientific research relating to climate change and, second, to formulate potential response strategies based on impact assessments (Luton, 2015). Hulme and Mahony (2010: 706) write that the IPCC is "central to the assessment, validation and mobilization" of climate change knowledge-claims. Beck and Mahony (2018) describe the IPCC as a "boundary organisation" which works at the intersection between science, society, and politics (see also Lahn and Sundqvist, 2017; Wynne, 2010). Here, IPCC acts as the global epistemic authority on climate change as the institution exercises a "significant influence on climate change knowledge, on public discourse about climate change and on climate policy development" (Hulme and Mahony, 2010: 713).

This significant influence constitutes a high amount of epistemic capital because the IPCC and connected epistemic authorities have the ability to make judgments on the credibility of research, influence non-scientific fields of society, and provide the scientific basis for important international agreements like the Paris Climate Agreement. In that context, Hughes (2015) argues—using Bourdieu's notion of capital and power—that the IPCC is the central actor within the global power-knowledge system of climate change, as an institution that has the power to authorize climate knowledge and to construct the meaning of climate change. Hughes (2015: 98) demonstrates that the IPCC's knowledge-claims have the power to profoundly influence concrete political action on a global scale as well as public debates in several fields of society, although she discusses the "symbolic power of knowledge" instead of epistemic capital.

Furthermore, Friedrichs (2011: 472) states that the IPCC attempts to maintain a "veneer in standard science," first, by emphasizing the importance of the peer-review process; second, by working with percentage values and confidence levels to avoid the utterance of absolute quantitative truth-claims; and third, by insisting on social and political neutrality. As the "strategic link between the science and the politics of climate change" (2011: 472), it is the intended purpose of the IPCC to provide a seemingly objective source of information for policy makers (Jebeile, 2020). Scientists involved in the IPCC slowly acknowledge the political and social implications of their research, but many "still rely heavily on an ontology that assumes a world in which climate science can be separated from climate policy and politics" (Luton, 2015: 157). In the common public understanding as well as in some academic studies, science and politics are frequently approached as two separate fields, where the IPCC is often uncritically perceived as a neutral and apolitical scientific organization (Lahn and Sundqvist, 2017). In that context, it is important to acknowledge that the IPCC "does not conduct its own scientific research and so does not directly create new knowledge" (Jebeile, 2020: 1); rather, the IPCC draws from existing research and combines it into synthesized reports and models.

The IPCC's publications and knowledge-claims are highly complex as they address the "interdependence of climate, ecosystems and biodiversity, and human societies" (IPCC, 2022: 7) by referring to a wide range of research and academic studies. But in the most basic sense, the IPCC (2013: 2) propagates the view that "human influence on the climate system is clear," that the global temperature increased due to an increase of greenhouse gases, and that anthropogenic causes "are *extremely likely* to have been the dominant cause of the observed warming since the mid-20th century" (2013: 4, italics in original). Moreover, the IPCC (2022: 11) writes that "human-induced climate change, including more frequent and intense extreme events, has caused widespread adverse impacts and related losses and damages to nature and people, beyond natural climate variability" and that "continued emission of greenhouse gases will cause further warming and long-lasting changes in all components of the climate system, increasing the likelihood of severe, pervasive and irreversible impacts for people and ecosystems" (2022: 8).

In short, what we consider to be the dominant, mainstream, and global knowledge on climate change, synthesized and communicated by the IPCC, argues that anthropogenic greenhouse gas emissions cause significant climate change and an increase in the global average temperatures that lead to adverse effects for the human population and natural environments. To mitigate the adverse effects of anthropogenic climate change, the IPCC suggests decreasing global greenhouse gas emissions as well as other adaptation measures like increasing human resiliency to global warming and natural disasters (e.g., changes in agriculture to ensure food production, building of dams to prepare coastal lines for rising sea levels). Here, the IPCC also monitors adaptation/mitigation measures and observes that "progress in adaptation planning and implementation has been observed across all sectors and regions, generating multiple benefits" (IPCC, 2022: 22).

Geopolitics and Epistemic Geographies of the IPCC Climate Knowledge

Regardless of the IPCC's "goal of maintaining objectivity" (Jebeile, 2020: 1), the synthesization and dissemination of climate knowledge by the IPCC are affected by several geopolitical and social epistemological dispositions. Relating to the IPCC, Demeritt (2001: 309) writes that "public representations of science seldom acknowledge the irreducibly social dimension of scientific knowledge and practice" since many scientists seem not to be aware of the political and societal impact of their work (Luton, 2015). Following the work of geographers like David Demeritt (2001), we argue that the IPCC's dominant climate change and environmental knowledge-claims are strongly influenced by global power relations, cultural traditions, and epistemic dispositions. The IPCC "is not value-free" (Jebeile, 2020: 1).

Nevertheless, since climate change is a highly sensible contemporary discourse, it is important to stress that "demystifying scientific knowledge and demonstrating the social relations its construction involves does not necessarily imply disbelief in either that knowledge or the phenomena it represents" (Demeritt, 2001: 310). Here, Wynne (2010) states that an analysis of the social and technical epistemology of anthropogenic climate change knowledge is not about denying anthropogenic climate change. Following this approach, we do not intend to discuss knowledge of anthropogenic climate change to undermine the truth of the dominant power-knowledge system of climate change nor to support climate skepticism. Rather, we examine all climate change knowledge-claims as socially constructed, regardless of if they originate from apocalyptic conspiracist environments or the IPCC. We engage with climate and environmental knowledge-making processes through a theoretical perspective in which "it has become a truism to claim that scientific knowledge is social knowledge" (Rolin, 2007: 115).

The IPCC's knowledge-making is mostly based on existing academic publications and literature from different sources, which are reviewed and synthesized by a variety of climate and environmental researchers. Relating to the selection of literature and researchers to work on the assessment reports, the IPCC is criticized for disciplinary, geographical, and gender biases (Pasgaard et al., 2015; Bjurström and Polk, 2011; Demeritt, 2001; Agrawala, 1998). The IPCC is dominated by white male scientists from Western, industrialized countries while lacking contributions from women, the social sciences, and experts from developing countries. This results in geographical, gendered, and disciplinary "knowledge hierarchies" (Mahony and Hulme, 2010: 708) within the IPCC. It is important to acknowledge that the "IPCC is an intergovernmental body under the auspices of the United Nations" (Jebeile, 2020: 7) so that existing and historically established geopolitical power relations of the UN affect the work of the IPCC. The interests of Global North states as well as the related "unequal distribution of both economic and epistemic resources can give rise to implicit biases" (2020) in the generation of knowledge that favors the Global North while neglecting the epistemic traditions, needs, and means of developing countries. These inequalities can result in tensions within the intergovernmental institution. For instance, Bert Bolin (2001), who was the first chairman of the IPCC between 1988 and 1997, writes that the allocation of a new chairperson prior to the preparations for the Fourth Assessment Report resulted in tensions between Western, industrialized countries and developing countries. The majority of African and Asian countries, but also the United States, backed the nomination of the Indian candidate Rajendra Pachauri, whereas most European and other industrialized countries proposed the re-election of the American Robert Watson. Eventually, Pachauri won the vote. The debate over the origin of the new chairperson illustrates geopolitical power tensions within the supposedly neutral knowledge-making institution. Here, Hughes (2015: 94) writes that since the establishment of the IPCC, "divergent attitudes between industrialized and developing countries" alongside geographical power differentials existed within the intergovernmental panel. Nonetheless, the vote of a chairperson from the Global South illustrates that the IPCC increasingly attempts to consider heterogeneous cultural perspectives on climate change (Agrawala, 1998) and, furthermore, that developing countries have the power to influence the organizational and knowledge-making structures of the IPCC (Hughes, 2015).

These examples of geographical power relations unveil *epistemic geographies* (Mahony and Hulme, 2018; Mahony, 2013) within the IPCC's knowledge-making. Epistemic geographies, however, are not only visible in the power relations between different scientists and states who contribute to the IPCC, but they are also at work in the scale of the knowledge produced. The IPCC and affiliated intergovernmental political institutions (e.g., UN, UNFCCC) introduce and disseminate global knowledge about the climate and natural environment (Mahony and Hulme, 2018; Jasanoff, 2013). The IPCC puts an emphasis on the formation of consensus forming by synthesizing existing academic research from different spatial origins. During that process of synthesizing and consensus forming, the local geographies and original researchers' epistemic standards under which the respective knowledge-claims were formed are taken into account to ensure the validity and reliability of the existing research (John, 2017).

However, in the presentation of the resulting knowledge, specific sites of local climate and environmental knowledge production and their respective influences on the formation of knowledge-claims are barely visible. In other words, distinct local physical geographies which fed the global climate models are absent in the final product of climate models as specific observations of small-scale weather and environmental processes are implemented into large-scale climate models. Hampel (2016: 245) writes that "scientists turn concrete particularities of place into universal exemplars of global climate change" to create global knowledge. Through "erasures of local specificity" (Jasanoff, 2010: 234), knowledge on the climate and environment are abstracted from the local social, cultural, physical and temporal realities from which the knowledges emerge from. While creating global knowledge of anthropogenic climate change, spatial environmental diversities are reduced to a "universal and global-scale problem of atmospheric emissions" (Demeritt, 2001: 312). Therefore, the IPCC does not only provide the dominant way of knowing about the climate and natural environment, but also a global way of knowing.

It is not only local physical indicators of a changing climate and environment that are neglected in the representation of global knowledge. A global way of knowing also lacks in local social and cultural meanings of climate, weather, and the environment (Hulme and Mahony, 2018; Jasanoff, 2010). Demeritt (2001: 316) writes that social meanings of climate and the environment "are ignored by scientists" as they aim toward the creation of objective and universal facts. Relating back to the geographical power relation within the IPCC, it must be noted that the definition of climate "in purely physical terms" and the separation of the cultural from the physical are especially dominant in Western ways of knowledge-making. In addition, Jasanoff (2010: 234) states that modern sciences such as climate science attempt to achieve "universality and heft" (Jasanoff, 2010: 234) by removing local sociocultural meanings and understandings of the climate and environment. With the local sites of knowledge creation erased, it is the climate model itself which is argued to be "the most powerful site of climatic knowledge production" (Hulme and Mahony, 2018: 399). This way of knowledge-making also results in a "dramatic simplification" (Demeritt, 2001: 316) of the complexity and diversity of the earth's climate and environment. Here, the production of a global and universal climate change knowledge needs to be understood as a process of homogenization and, in particular, as an abstraction of local geographies.

Post-Normal Science

The IPCC's knowledge, as well as modern climate science more broadly, is largely based on epistemic strategies which emphasize quantitative methods. Historical and contemporary temperatures, greenhouse gas records and other types of observational and modeled data are used to feed climate models and to empirically test and confirm anthropogenic climate change (Hampel, 2016; Kappas, 2009). In that context, the IPCC's methods require a high degree of trust toward previous knowledge, a fundamental concept in knowledge creation (Shapin, 1994). The IPCC attempts to ensure the quality of previous research through critical assessments and reviews of previous research that is synthesized in the IPCC reports (John, 2017). In addition to the physical dimensions of climate and environmental change, the IPCC reports also include assessments of social impacts and possible mitigation strategies, although these sections are dominated by quantitative economics (Bjurström and Polk, 2011). The epistemic strategies and scientific standards approved, employed, and disseminated by the IPCC can be conceptualized as the dominant episteme on climate change and environmental knowledge. However, the IPCC's knowledge-producing practices and other practices of quantitative climate science are often described as *post-normal science* (Luton, 2015; Turnpenny, Jones, and Lorenzoni, 2011; Ravetz, 2011; Hulme, 2009) as they challenge the idealized imagination of objective, already-observed truths and political neutrality of applied sciences as a form of *normal science*.

Ravetz (1986: 422) describes post-normal science as a form of knowledge-making in which "facts are uncertain, values in dispute, stakes high and decisions urgent." As in the IPCC reports, these characteristics can be identified in mainstream climate and environmental sciences. Post-normal science is a conceptualization of scientific practice which acts as a response to what is described as "normal science," a term coined by the English philosopher Thomas Kuhn (2012 [1962]). Kuhn (2012 [1962]: 10) defines normal science as "research firmly based upon one or more past scientific achievements, achievements that some particular scientific community acknowledges for a time as supplying the foundation for further practice." Normal science is "the science done when the fundamentals stand beyond question" (Sharrock and Read, 2002: 37). It is practiced under the assumption that there must be possible solutions to certain problems. Hence, normal science is described by Kuhn as a puzzle-solving practice which aims toward the completion of a puzzle through the production of knowledge. In addition, Sharrock (2002: 39) writes that practitioners of normal science will "set aside" questions which seemingly "cannot be solved."

Ravetz's (1986) characteristics of post-normal science (I: uncertain facts, II: disputed values, III: high stakes and urgent decisions) contradict Kuhn's notion of normal science. First (I), and relating to the uncertainty of climate change facts, Hulme (2009: 107), an advocate of the conceptualization of post-normal science, argues that climate change knowledge "will always be incomplete, and it will always be uncertain." This uncertainty is largely based on the complexity of, and connection between, the earth's ecosystems and spheres, for instance, apparent in interconnected ecosystems, carbon cycles, or rebound effects (Kappas, 2009). Furthermore, specific quantifications of how the global climate system responds to factors such as greenhouse gases or radiative forces (climate sensitivity) remain uncertain and, are therefore, a matter of dispute between scientists (Katzav and Parker, 2018). Kuhn's imagination of normal science as a "textbook style of investigation becomes less effective" (Ravetz, 2011: 150) in the case of climate change due to the complexity of the earth's systems and the human impact on nature and climate. Climate science constitutes, to use Kuhn's language, a puzzle which cannot be solved. Traditional guiding principles of the idealized idea of normal science, which pursues the discovery of objective truth and factual knowledge, or completion of a puzzle, have to be adjusted in post-normal science (Ravetz, 1999). The aim of climate and environmental sciences is to increase the understanding of the earth's ecosystems and climates to approximately predict future climates by developing a range of scenarios. In Kuhn's understanding, a paradigm which sets the fundamentals is largely absent in climate science since there are multiple different, partially concurring, epistemic strategies employed simultaneously within mainstream climate science, for instance, different climate models that calculate the expected climate and environmental change using distinct methods (Bray and Martinez, 2015).

Second (II), and as a consequence of the uncertainty within climate science, the values of climate change (for instance, the predicted increase in global temperature, increase of sea levels etc.) are under dispute. The IPCC develops a range of future scenarios and employs a "consensus mode of establishing and communicating knowledge" (Hulme, 2009: 88). Climate change knowledge is, thus, largely a result of social processes, consensus forming, negotiation, and cooperation between scientists and other experts. The consensus model "has been a source of both strengths and vulnerability for the IPCC" (Hulme and Mahony, 2010: 711) because the collective judgment on the uncertain knowledges increases the quality and robustness of climate science, while deviant voices are at risk of being marginalized. Consensus, however, cannot establish an objective fact, a truth in an absolute or a strict sense of knowledge because it has to be understood as a blend of a number of separately produced subjective truths. Here, the IPCC's methodology of consensus reminds of Shapin's (1994: 5) studies of knowledges in which he argues that "whatever bears the marks of collective production cannot be truth and honored as such."

Third (III), climate science, as practiced by the IPCC, deals with high stakes and urgent decisions. The adverse effects of climate change and the resulting environmental change threaten human and non-human life, potentially increase the strength of natural disasters and, furthermore, endanger water and food supplies on global and regional scales. Thereby, climate change becomes even more a concern of security (Dalby, 2013). Due to these adverse effects of a changing climate, a prompt response by political decision-makers on a global scale is required. Climate knowledge, in a post-normal understanding, is intended to work as a force for political action and social change (Turnpenny, 2012). As mentioned earlier, the IPCC works as a boundary organization between science and politics to bring about a prompt political response to climate and environmental change on a global scale. The political need for climate and environmental knowledge is described as a "post-normal situation of policy relevant science" (Luton, 2015: 158).

The concept of post-normal science itself is a matter of discussion and a focus of criticism (Turnpenny, 2012). For instance, the terminology of "post"-normal conveys the impression that post-normal science "represents an epistemological change" (Turnpenny, Jones, and Lorenzoni, 2011: 295, following Weingart, 1997) from normal to post-normal science. Here, the concept of post-normal science neglects that knowledge-making practices were always informed by social processes while facing uncertainty and political needs. In that context, Turnpenny, Jones, and Lorenzoni (2011) argue that science has always been social, but that the notion of post-normal science in contrast to an idealized imagination of objective science highlights the always present social features of knowledge-making. Hence, post-normal science should not be understood as a distinguished way of knowledge-making or as a new methodological paradigm, but rather as a theoretical concept which highlights the social characteristics and uncertainties in knowledge-making processes.

In relation to the societal character of knowledge-making, and regardless of the IPCC, it must be noted that climate and environmental knowledge-making is not merely an academic scientific enterprise but a form of knowledge which arises out of scientific as well as societal processes and knowledges (Turnpenny, Jones, and Lorenzoni, 2011). Ravetz (1999: 651) explains that "extended peer communities," which exceed the academic scientific community, contribute "extended facts" (1999) to debates, for instance, in the form of local and indigenous knowledges on sustainability "in ways for which the accredited experts, with the best will in the world, are not prepared" (1999).

Overall, the IPCC's climate science supports the conceptualization of climate and environmental knowledge-making as a collective good, an accomplishment and a social construct. The close ties between climate science and politics illustrate that climate knowledge cannot be discussed separately from society and politics. Furthermore, the epistemic strategies of the IPCC, which are frequently described as post-normal, show that climate and environmental knowledges face uncertainties and are a result of a social process of consensus building. A thorough, all-encompassing, absolute, and objective truth on climate and environmental change is not produced by the IPCC. Therefore, climate and environmental knowledge fit Shapin's (1994) understanding of truth and knowledge of a social institution. Still, extremely high probabilities show that anthropogenic climate change very likely exists—a thesis which is supported by the IPCC's epistemic community.

Conclusion: Dominant Knowledge and Counter-Knowledge

This chapter argues for a rather social constructionist approach to the production of truths, science, and valid knowledge in which the definition of an objective fact/truth is always a subjective judgment that depends on space, time, cultural histories, and social environments. Due to that subjectivity, the historical, geographical, and sociological dimensions of environmental sciences, including climate science, result in a plurality and diversity of climate change epistemes and resulting knowledge-claims. This plurality becomes evident when looking, for instance, at academic research which investigates how climate change knowledges are influenced by history and geography (Mahony and Hulme, 2018; Hulme, 2011; Brace and Geoghegan, 2010; Livingstone, 2002a), gender (Pearse, 2017), religion (Hulme, 2015a, 2015b; Bergmann, 2009; Gottlieb, 2006, 2004) or political attitudes (Lewandowsky, Ecker, and Cook, 2017; Lewandowsky, Oberauer, and Gignac, 2013; Arbuckle, 2017).

We accept the IPCC's knowledge as true due to our personal acceptance dispositions that are informed by our cultural background and education, that taught us to assign a high level of authority toward the quantitative methods that are employed by a large number of scientists that have a university degree. Therefore, we trust the external knowledge communicated by the IPCC, in particular because our internal ways of acquiring knowledge about a thing as complex as the climate are rather limited. But this trust is not ubiquitous. Other social or religious movements reject intergovernmental scientific institutions and mainstream climate science more broadly because they do not trust the IPCC and other epistemic authorities. Based on negative identity prejudices and trust in contradicting epistemic strategies (like trusting the Bible), certain religious and cultural movements produce alternative knowledge that counters the knowledge-claims and epistemic strategies of the dominant IPCC.

Climate change, framed as a societal and scientific discourse, can be described as a "battleground" between "different ways of knowing" (Hulme, 2009: xxvii). Here, climate change further functions as a "proxy" (Turnpenny, 2012: 403) for other societal conflicts since "clashes over science are often the surface manifestation of deeper political or cultural conflicts" (Jasanoff, 1997: 582). Due to the conflict between different ways of knowing and the significant influence of societal factors, history, and geography on the formation of climate knowledges, we argue that it is appropriate to discuss climate change knowledges as what Goldman (2002) would describe as a loose sense of knowledge, as socially constructed knowledge, or a social institution.

Yet, different ways of knowing about climate change are not just out there; they compete for power on different societal scales, resulting in different power-knowledge hierarchies. On a global scale, the IPCC's way of knowing about climate change can be approached as the dominant knowledge as it affects not only the generation of global climate change policies but also several national, regional, and local adaptation and mitigation measures. Therefore, knowledge-claims which teach that anthropogenic climate change is not real or not an issue of any relevance are conceptualized as counter-knowledge in this book. Still, it is always an interplay between time, scale and space that decides what constitutes the dominant truth and what is the counter-knowledge. In some social environments, knowledge of anthropogenic climate change is rejected by the institutions in power that govern the general discourse in this particular space.

But what matters in our research on apocalyptic conspiracism is not just the "objective" power-knowledge hierarchy in terms of which beliefs have a higher impact on political decisions and public opinions, but the subjective power-knowledge hierarchy of the individuals who feel neglected or oppressed. For apocalyptic conspiracists, it does not matter who is actually in power, as these individuals or social groups construct themselves as suppressed by institutions in power, regardless of their actual epistemic capital. For instance, independently from Trump's withdrawal from the Paris Climate Agreement, the American apocalyptic conspiracists we examine in this book still portray themselves as endangered by, and victims of, evil elites, may it be the globalists of the anti-Christian UN or the minions of the American Deep State.

3

Analyzing Digital Knowledge Discourses/Spaces

This chapter outlines and develops this book's methodological approach and data collection within the spaces in which the dominant global knowledge-power system of anthropogenic climate change is contested and countered. The internet, as a "discursive site of practice" (Barr, 2011: 15), is of particular interest for our data collection since existing research shows that digital environments constitute fertile grounds for developing, distributing, and discussing apocalyptic as well as conspiracist discourses (Robertson, 2018, 2016; Knowles, 2013; Barkun, 2013; Howard, 2011, 2010, 2006; Dittmer, 2010). Sharman (2014: 161) writes that online environments constitute "new sites of knowledge production" in regard to climate change counter-knowledge that demand increased academic engagement. Moreover, the internet is a pivotal site of counter-knowledge creation, distribution, and consumption concerning conspiracist discourses (Ahmed et al., 2020; Duplaga, 2020). Due to the importance of the internet in enabling the powerful "contagious conspiracism" (Sturm and Albrecht, 2021a) of the Covid pandemic, academic research on conspiracism and the internet increased notably in recent years. But as Mahl, Schäfer, and Zeng (2022) note, much of this research within the humanities lack theoretical frameworks that consider the particular features of conspiracy discourses in online environments.

We suggest a framework that combines the conceptualization of apocalyptic conspiracist discourses as counter-knowledge with ideas of the geographies of knowledge. In this chapter, we argue for an approach that looks at the internet and its different (web)sites as "intellectual, social and cultural arenas" (Livingstone, 2003: 6) in which counter-knowledge is created, distributed, and consumed. As the rise of online conspiracism in recent years has shown, it is important to adapt established social epistemological concepts to the internet as digital technologies profoundly change the geographies of knowledge production (McLean, 2020; Lee, 2015; Graham, 2014; Livingstone, 2003) and this chapter should be understood as a basis for considerations in a space of inquiry that we call the *digital geographies of knowledge*.

In the first section of this chapter, we provide a methodological foundation for the analysis of discourses that emerge out of the internet by applying social epistemology to digital spaces. We argue that the internet should be conceptualized as digital space or a network of digital spaces, each with a distinct cultural history, respective conceptions of the truth and accepted epistemic strategies. Some of these spaces can have unique characteristics that allow the dissemination of counter-knowledge as they exist outside the traditional epistemic power of mainstream media outlets and epistemic authorities. Yet, these digital spaces do not exist separately from the offline world, as historically formed dispositions, cultural histories, and epistemic strategies continue to exist on the internet and they influence how knowledge is formed and consumed.

The second section of this chapter describes the collection of primary data from the internet. We engage with digital spaces in which (potential) apocalyptic conspiracists conduct "their own research" to disprove what is believed by the mainstream. The internet provides a vast amount of textual, visual, and audio material on climate change and geopolitics, which allow us to piece together how apocalyptic conspiracist ways of knowing function. Through critical discourse analysis, we examine the underlying acceptance dispositions, epistemic strategies, and cultural histories that affect the rejection of dominant knowledge-claims on climate change. In that context, the discursive construction and justification of apocalyptic conspiracist counter-knowledge are of particular interest as apocalyptic conspiracists are seemingly successful in convincing increasing shares of the population in several Western countries of their respective way of knowing about the world.

Section I: Digital Geographies of Knowledge

The assemblage between space, place, science, and knowledge is an established area of inquiry in human geography and science and technology studies. Researchers of the *geographies of knowledge* specifically ask how physical, social, and cultural spaces influence access to knowledge, the perception of knowledge, as well as the knowledge production more broadly. In addition, the relationship between knowledge and social power relations, from a local scale up to a global or geopolitical scale, is frequently discussed in geographical studies of knowledge (Gregory, Meusburger, and Suarsana, 2015). An underlying assumption of studies within the geography of knowledge is that imaginations of the truth differ depending on geography and that, in short, spaces matter in the construction, dissemination, and perception of knowledge (Livingstone, 2005a, 2003).

The geographies of knowledge engage with the physical properties of a certain space, like a laboratory that allows certain scientific practices or ways of creating knowledge that are not possible in other spaces, for instance, due to specific equipment. But the geographies of knowledge further explore the social and cultural features of spaces, their respective epistemic preferences, and forms of human interaction that affect scientific practice, perceptions of the truth, and ways of knowing about the world. The scale of research is wide, from colleagues in a workplace, or entire historically formed epistemes on a national or large-scale cultural level.

In short, what is believed to be the truth, and what is perceived as a legitimate way of creating knowledge, depends on geography. Here, Livingstone (2002a: 95) explains that the designation of something as the truth changes "over time and from place to place, and there are unquestionably social histories and cultural geographies of what people have considered the appropriate practices you have to perform, or what epistemic obligations you have to fulfil, in order to arrive at something called 'truth'."

In the last few decades, new spaces and places have emerged due to new computer technologies. The specific characteristics of these spaces and places influence knowledge in a variety of ways—spaces on the internet. Here, it is important to recognize that the internet is "not an amorphous, spaceless and placeless cloud" (Graham, 2014: 99), but a complex network of interactions which accommodates a multitude of spaces in which knowledge-claims are produced and contested. Our argument in what follows is that the spaces enabled by digital technologies have, like all other traditional spaces, specific ways of knowing about the world that depend on social histories and cultural geographies of that particular space. In that context, we follow Graham (1998) and argue that what might be conceptualized as the "online" versus "offline" are not binary or demarcated spaces. Rather, we will demonstrate that online and offline spaces do not exist separately from each other but that both need to be understood as a complex, mutually affecting set of social relations (Graham, 1998; Massey, 1993) and "constitutive of systems of human interaction" (Livingstone, 2003: 7), which transcend any binary conceptualizations. Therefore, the geographical metaphor of a cyberspace will not be used in this research. The cyberspace metaphor might help to "make tangible the enormously complex and arcane technological systems which underpin the internet" (Graham, 1998: 116). Still, the imagination of a cyberspace neglects the "pervasiveness of digital technologies" (Ash, Kitchin, and Leszczynski, 2018: 26) in the "spaces and practices of everyday life" (2018: 32). A similar argument is made by Snee et al. (2016: 4), who argue that the internet and social networking platforms are "embedded in multiple contexts of everyday life" and therefore cannot be conceptualized separately from physical, offline spaces. Historically established cultural beliefs, social geographies, and epistemic preferences get transferred to the internet, while, conversely, knowledge developed and acquired on the internet affects individual and societal practices and beliefs due to the "deepening reach of the digital in everyday life" (McLean, 2020: 34). Practices in digital spheres alter public societal discourses and shape political decision-making processes (Pinkerton and Benwell, 2014; Dahlgren, 2005).

Here, the terminology which allows us to exceed any binary geographical imaginations of the online and offline is the notion of *the digital*. In their discussion of a digital turn within the discipline of geography, Ash, Kitchin, and Leszczynski (2018) develop, based on previous work by Lunenfeld (1999), a geographical notion of the digital which dissolves taken-for-granted boundaries between the online and the offline. The digital, for Ash, Kitchin, and Leszczynski (2018) include all discourses, productions, and experiences of space, place, nature, and environments which are reshaped by new communication technologies, regardless of whether they are taking place on a computer screen or in any physical space. The digital is a "genre of socio-techno-cultural productions, artefacts and orderings of everyday life that result from our spatial engagement with digital mediums" (2018: 26). Applying this notion of the digital to the understanding of space as a set of social relations, a *digital space* can be defined as a set of social relations which are accessed, created, transformed, or affected by digital technologies.

A digital space can be a website, a YouTube comment section, a Meta (formerly Facebook) group, an X (formerly Twitter) page, or an online forum discussion board like 4chan and 8chan. A digital space is any "intellectual, social and cultural arena" (Livingstone, 2003: 6) that can be accessed through digital technologies or that emerges due to digital technologies. McLean (2020: 34) uses "the 'more-than-real'" as a concept that "explains the paradoxical ways that digital spaces amplify and collapse geographies, reworking spatial connections and disconnections." The internet and digital spaces should not be approached as "unreal," but "more-than-real" and as spaces that "can challenge normative spatial relations, sometimes in surprising ways" (2020: 9).

Due to this impact of the digital on different human and social geographies, Ash, Kitchin, and Leszczynski (2018: 35) suggest we "think about how the digital reshapes many geographies" and, for our purposes, how the digital reshapes the geographies of knowledge. In short, our argument is that if the internet accommodates digital spaces (McLean, 2020; Ash, Kitchin, and Leszczynski, 2018; Graham, 2014, 1998), and if "spaces matter" (Livingstone, 2005a: 100) in the construction, circulation, and consumption of knowledge, then digital spaces are epistemologically relevant, and they influence perceptions of the truth and therefore how people know. Incorporating digital spaces into the field of geographies of knowledge allows us to expand the field to form *digital geographies of knowledge*. Those digital geographies of knowledge apply general questions of social epistemology, science and technology studies, and the traditional geographies of knowledge to digital spaces, which, like other spaces, consist of a network of social relations which influence the knowledge production, dissemination, and perception. Here, different digital spaces possess different epistemes that govern which knowledge-creating strategies are considered as legitimate and valid forms of knowledge-generation. While Bible study might be an authoritative epistemic strategy in American evangelical online forum discussion boards, it is a less legitimate source of knowledge-generation relative to "secular" digital spaces.

Liberalization of Knowledge in Digital Spaces

Discussing digital spaces as a set of social relations, like any other space, allows us to apply a number of theoretical concepts discussed in the previous chapter to digital spaces. Shapin (1994) and Foucault (1980) draw attention to the hierarchical order of knowledge-creating systems and other power relations which are inscribed in place and geography. Graham (1998: 181) argues that digital spaces constitute an arena for "complex social power struggles," which are not separated from power struggles in the public sphere and physical world. As was made clear by, for instance, the Russian interference in the 2016 US presidential election via the distribution of misinformation about Hillary Clinton (Persily, 2017; Maréchal, 2016), the dissemination of knowledge in digital spaces can change political realities at different scales. Yet, power relations and power hierarchies are structured differently in digital spaces compared to traditional physical public and social spaces. A society's regime of truth and established hierarchy of knowledge previously instantiated by traditional mainstream media, epistemic authorities such as universities and political institutions in power are challenged as a wider spectrum of beliefs, knowledge-claims, epistemic dispositions, and previously marginalized cultural traditions enter the public sphere through digital technologies (Barkun, 2013; Morris and Morris, 2013; Kaufhold, Valenzuela, and De Zúñiga, 2010; Baum and Groeling, 2008).

Social media platforms or alternative news websites are examples of the contemporary changing power and knowledge hierarchies enabled by digital technologies and their spaces. While the trust in the traditional mainstream media is declining in the United States (Persily, 2017), the power of alternative and social media platforms on the internet, such as Breitbart or Infowars, as well as news-sharing-themed URL banners like Facebook or Twitter, is increasing. These (web)sites can "bypass" traditional institutions in power of the public discourse as "media outlets that challenge dominant political, economic and media power" (Lee, 2015: 320). On the internet, individuals or social groups can "deconstruct official versions of the truth" and offer alternative truths or counter-knowledge at the same time (Aupers, 2012: 27).

Certain actors or groups might have less epistemic capital in the physical world and traditional mainstream media, but a high amount of epistemic capital in the digital world. Alex Jones, for example, started his broadcasting career in radio and TV, but he only became an influential political figure outside of traditional mass media after he was fired from the radio station KJFK-FM in 1999. Through his digital alternative media outlet Infowars, Alex Jones successfully pushed a counter-hegemonic apocalyptic conspiracist and populist nationalist discourse—which disseminates the belief in a war against global elites who allegedly suppress American people—into the American mainstream (Van den Bulck and Hyzen, 2020; Robertson, 2015).

Morris and Morris (2013: 589) explain that "internet created the possibility of a forum for any single person or group to be heard by all" because the costs of creating, distributing, and accessing any form of content are lowered by digital technologies. From geographical perspective, it is highly relevant that digital technologies allow for knowledge-claims and beliefs to be shared in digital spaces which exceed a single person's or community's social and cultural environment and capital in the physical world (Lee, 2015; Graham, 1998). Nonetheless, it is not just the ability by a single person or group to reach a wider, potentially global audience, but also the "volume and speed of knowledge flow through the internet" (Hsu, 2015: 467). The website of Alex Jones' Infowars offers alternative framings of current events with the same speed and topicality, and in the same visual format, as mainstream news websites. Such sites are not just "alternative" media but rather the first source they encounter to make sense of events. Yet, established acceptance dispositions that influence the perception of a communicator of knowledge depend on the respective cultural histories and social geographies of a digital space. In short, what is accepted as true on Infowars might still be perceived as false on the *New York Times* website.

However, while the audience of alternative media and their accounts of the truth are still relatively small compared to traditional mass media, the "influence on the public opinion process can be disproportional to its size" (Lee, 2015: 336). The audience of traditional mainstream media tends to be less politically active than users or creators of alternative media (Kaufhold, Valenzuela, and De Zúñiga, 2010). The storming of the American Capitol Building in 2021 illustrates that a relatively small number of people who organized the insurgency on the internet, driven by QAnon ideologies and other conspiracist beliefs, drew worldwide media attention as they profoundly affected the public political sphere in the United States.

In sum, digital spaces are more horizontal social spaces of participation in which the means of knowledge production and communication are more equally distributed compared to traditional physical social, cultural, and political spaces. This feature of the internet causes an overall liberalization of knowledge since more and more people can acquire and distribute knowledge online. Societal imbalances between the creators and consumers of knowledge are less pronounced in digital spaces. In that context, and as Morris and Morris (2013) argue in the political, social, and cultural context of the United States, the internet further helps to minimize the knowledge and participation gaps between low and high socioeconomic status citizens. Therefore, alternative ways of knowing and counter-knowledge can be established as true in digital spaces, which then compete for power in society as a whole.

Digital Spaces and Power

A special feature of digital spaces that affects the digital geographies of knowledge is the interaction between humans and "more-than-human actors," for instance, artificial intelligence, bots, and algorithms that can govern social practices in digital spaces (Amoore, 2020; McLean, 2020). Digital spaces are "transforming social relations" (Ash, Kitchin, and Leszczynski, 2018: 29) which applies to social interactions between humans but also social interactions between humans and more-than-human actors (e.g., bots or algorithms that regulate or generate content on the internet). McLean (2020: 26) writes that "digital geographies are co-produced by relations between people, technologies and the more-than-human." Here, algorithms on social networking sites like Twitter or YouTube have a profound impact on the information human users are exposed to. In a competition for attention, personal data, and time spent in a certain digital space, social media platforms and their algorithms select content based on the users' previous search history and preferences, resulting in filter bubbles or echo chambers that influence the opinion-forming and acquisition of knowledge in digital spaces (Geschke, Lorenz, and Holtz, 2019). An echo chamber is a social space in which people expose themselves to an echo of their preferred ideology that fortifies and reinforces preexisting beliefs through collective reaffirmation (Törnberg, 2018). Within digital echo chambers, concurring ideologies and ways of knowing are filtered out, and consequently, individuals are only exposed to information which seems to correspond with preexisting beliefs. In echo chambers, people are likely to be exposed to knowledge that matches their dispositions. Although "the power to expose oneself to perspectives from the other side in social media lies first and foremost with individuals" (Bakshy, Messing, and Adamic, 2015: 1132), digital echo chambers and filter bubbles contribute to an epistemic fragmentation of societies (Geschke, Lorenz, and Holtz, 2019). These more-than-human actors, the algorithms, and bots that create an echo chamber have a profound impact on societal opinion-forming. Individuals on the internet perceive certain knowledge as more socially accepted than it is when they are just exposed to knowledge that corresponds with previously held beliefs, where socially accepted belief constitutes true knowledge (Shapin, 1994, see Chapter 2). Still, physical spaces, their cultural histories, and social geographies remain important in the creation, dissemination, and acquisition of knowledge. It is important to stress that echo chambers, too, are relational. Bastos, Mercea, and Baronchelli (2018: 13) explain that "echo chambers are connected with homophilous dependencies in physical social networks," which might just "spill-over" (2018) to digital social networks.

Regardless of the apparent liberating and democratizing character of the internet, which allows the establishment of concurring knowledge-power systems in the public sphere, it must be acknowledged that control, surveillance, and other forms of power do exist in digital spaces. The internet is not a completely anarchistic environment free of any authority, gatekeepers, and governmentality, nor is it an ethereal spaceless network; it is made up of nodes and networks of nodes which upload this information from somewhere. In digital spaces, power might be more equally distributed relative to traditional social spaces and physical environments, but governments, private companies, web hosts, administrators, and algorithms have the ability to track, control, and shape the behavior of the internet's users. In contrast to conventional physical social spaces, "power is exerted subtly through distributed protocols that define and regulate access to resources and spaces and reshape behaviour" (Ash, Kitchin, and Leszczynski, 2018: 31).

YouTube's watch-next algorithm has contributed to a conspiracy boom and the early popularization of QAnon in 2017 through automated recommendations of conspiracist contents, including end-of-the-world prophecies (de Zeeuw et al., 2020; Faddoul, Chaslot, and Farid, 2020). Due to the rising power of conspiracism and scientific counter-knowledge in the public sphere, epistemic authorities, national governments, and some social media platforms slowly began to attempt to regulate knowledge-containing discourses on the internet. The most prominent example of this might be the ban of Alex Jones on several social media platforms like YouTube, Twitter/X, or Apple Podcasts. Facebook, for example, said that they banned Jones' and Infowars' contents due to "glorifying violence, which violates our graphic violence policy, and using dehumanizing language to describe people who are transgender, Muslims and immigrants, which violates our hate speech policies" (Coaston, 2018).

Still, several users of the internet, not necessarily conspiracists or advocates of counter-knowledge, criticize this increasing regulation of digital spaces as they see the free speech on the internet under attack. Allegedly "championing free speech" (Van den Bulck and Hyzen, 2020: 47), popular social media platforms like Reddit, 4chan, and 8chan welcomed "communicators banned elsewhere" (2020). Of course, not all platforms follow the same policies concerning the regulation of supposed misinformation and conspiracism, which shows that the power relation between users and digital or societal authorities depends on the particular digital space. It is mostly large and popular social media platforms like Facebook, Twitter/X, and YouTube, as well as the comment sections of large news outlets, that regulate discourses. In contrast, independently hosted websites (like infowars.com) can continue to self-govern their digital spaces. The regulation of content in digital spaces somewhat confirms the conspiracist worldview since "content removals and bans may also have the unintended effect of strengthening the beliefs of conspiracy theorists for whom such interventions are proof that they are in the process of uncovering deeper secrets" (Bruns, Harrington, and Hurcombe, 2020: 26). Other practices, like *Instagram's* information banners that get automatically attached to any posts that mention terms like "vaccine" or "corona," can also confirm a conspiracist worldview.

YouTube also attaches information panels to potentially conspiracist videos to provide "fact checks" (YouTube Help, no date), but "information panels are only available in a limited number of countries/regions and languages. We're working to bring information panels to more countries/regions" (no date). In our research, we noticed that YouTube shows an information panel about climate change, including a link to the UN's definition of climate change, when accessing videos from EndTimeMinistries (EndTimeInc, 2019a) on YouTube from Germany. But when we accessed the same link/video with an American IP address, there is no panel or fact-checking on the website because "in the United States, we [YouTube] only show fact checks from publishers based in the United States." The different presentations of the same video, depending on the user's geographical location, again show that space matters when consuming knowledge on the internet. While users from the United States only encounter the particular knowledge-claims of *EndTimeMinistries*, users from other countries automatically get exposed to a concurring knowledge-claim and the dominant knowledge on climate change. In general, the practice of providing fact checks, banners, or information panels could suggest for adherents of conspiracism that powerful actors of the digital sphere like Instagram or YouTube attempt to enforce a certain way of knowing about climate change or Covid.

However, due to the deleting of fake news, misinformation, and conspiracist content on the internet, some conspiracist authors and organizations created their own digital spaces outside of the regulatory power of social media companies. Infowars started their own video platform, Unbanned, to distribute their conspiracist content, primarily Alex Jones' talk shows, and other videos that distribute counter-knowledge on important topics such as geopolitics, climate change, Covid, or American politics more broadly. After Donald Trump was banned from Twitter/X in 2021 due to his involvement in the storming of the American Capitol, the former American president announced his own social media platform TRUTH Social, a "social media platform that encourages an open, free, and honest global conversation without discriminating against political ideology" (Truth Social, 2022).

Section II: Analyzing Contemporary Apocalyptic Thought in Digital Spaces

The specific characteristics of the internet alter religious beliefs and practices, including those of American evangelical apocalypticism (Wilson, 2017; Dittmer, 2010; Howard, 2006). Beesley (2011: 45) argues that the "internet has completely transformed the apocalyptic discourse" and that the particular geographies of the internet contribute to this profound alteration of the field of American apocalypticism. In that context, Heidi A. Campbell (2017: 16), who notably contributed to the theorization of the academic field of *digital religion studies*, envisions the internet as a unique "social context and space where culture is made and negotiated." Campbell (2017: 17) also dismisses a binary conceptualization of the offline and online but argues that digital technologies enable "hybrid spaces of practice" that transcend conceptualizations of "online" and "offline." While these terms can provide important categories of analysis, they cannot live up to the complex interdependencies of the digital and the physical world. Therefore, and similar to our conceptualization of the digital outlined above, digital religion studies holds that the internet should not be treated as a self-enclosed spatial entity (Tsuria et al., 2017). Building on this notion of the digital from the field of digital religion studies, we will approach apocalyptic conspiracism on the internet as a discursive field that emerges out of American apocalyptic histories while considering the distinct features of the internet and their impact on evangelical apocalypticism.

However, addressing American evangelicalism and the social dimension of religious digital spaces, Howard (2011, 2010) explains that the internet accommodates "virtual ekklesia." These are religious communities in digital environments that connect believers from different geographical locations. In that context, a sense of identity and community is not created by physical locations or geographical origins but by sharing the same beliefs. Knowledge is the decisive factor that makes people feel like part of a community, as "individuals must be marked as insiders who share their special knowledge, while those who do not share this knowledge must be marked as outsiders" (Howard, 2011: 14). Hence, social groups on the internet not only determine what is to be considered as true knowledge through social negotiations and discursive practices (see Chapter 2), but the social group itself is defined through a shared conception of true knowledge. This can lead to a homogenization of beliefs as well as a lack of debate and criticism within a virtual ekklesia. Opposing knowledge-claims can result in an exclusion from the digital community, in particular in combination with the active gatekeeping practices in apocalyptic digital spaces (Knowles, 2013).

While individual digital spaces, websites, and their ekklesia might not provide a variety of contrasting knowledge-claims due to the interdependencies between knowledge and group identity, the different End-Times beliefs on the internet as a whole are multifarious. An important condition that causes the increasing diversity of apocalyptic thought is the lack of institutional authority or gatekeeping on the internet. Within digital spaces, American prophecy writers can act not only outside of epistemic constraints set by the state, media, and other epistemic authorities (with the limitations addressed in the previous section) but also independently from traditional religious authorities. Beesly (2011: 50) argues that digital technologies moved religious apocalypticism away from any institutional control and "one step closer to individualism and egalitarianism." The power of an apocalyptic authority on the internet can even exceed the power of traditional religious authorities when digital prophecy writers acquire a followership that exceeds the social influence of a preacher in a church (Howard, 2011; Beesly, 2011). Yet digital technologies should not be understood as the decisive factor in the diversification of American apocalypticism but as further accelerating the religious individualization of American evangelicalism that persisted for centuries (Beesly, 2011; Frazier-Crawford Boerl and Perkins, 2011; Howard, 2006; Marsden, 1980; see Chapter 4). Evangelical tradition of "ground-up eschatology" further motivates a large number of people to share their respective imaginations of the truth online (Knowles, 2013: 130).

Exploring the reasons for the diversification of religious belief in the digital age, McClure (2017: 486) argues that digital technologies encourage us to *tinker* around with religious beliefs, as he describes the internet as a "prime carrier" for religious pluralism. The internet allows us to access knowledge about a variety of different belief systems and traditions, which exceeds the ones accessible in non-digital social environments. This variety of truths in digital spaces, according to McClure, can affect individuals to question their previously formed reality and, hence, to tinker with a new belief system through an individual combination of beliefs and discourses found online and offline. McClure's notion of tinkering suits the apocalyptic practice of "combin[ing] ideas of varying religious, political, social, economic and cultural persuasions into single depictions of how the world will end" (Beesly, 2011: 69). Similarly, Barkun (2013) states that improvisational millennialism requires a lack of gatekeeping and a variety of easily accessible previous ideas/concepts which can be used to construct novel prophetic ideas. Such conditions are ever-present on the internet. In that context, improvisational millennialists "relentless and seemingly indiscriminate[ly]" (2013: 18) borrow from a wide range of ideological traditions like esotericism, astrology, homeopathy, or religious knowledge and "combine elements so disparate that it is often impossible to determine what if any influence predominates" (2013: 22).

Within American evangelicalism, such apocalyptic improvising and tinkering on the internet is also grounded in the American apocalyptic tradition and its "profound narrative plasticity" (Howard, 2006: 26), which makes apocalyptic Christian rhetoric "highly sustainable" (2006) in online environments. Howard (2006) argues that apocalyptic discourses are generally sustainable (in terms of being durable, long-lasting, always present, and not environmentally friendly) because the overall narrative of an apocalypse allows for an "infinite variety of minor revisions," adaptations, and the introduction of new ideas. This remains, however, somewhat paradoxical given that biblical authority should prescribe a relatively static narrative (at least in rather conservative Christian apocalyptic narratives) whose "truth is beyond question" (2006: 28). But this truth is broadly defined, and what constitutes the truth always depends on the social environment and cultural history of the particular space in which a perception of the truth is formed (Livingstone, 2002a; Shapin, 1994). In this milieu, evangelical apocalyptic prophecy writers have been able to produce an exceptional variety of apocalyptic knowledge-claims without contradicting the core principles of evangelical biblicism.

Critical Discourse Analysis and Data Collection

Methodologically, we analyze the construction of American End-Times beliefs, which hold that global crises, specifically climate change, must be understood within the framing of an apocalyptic superconspiracy. We explore how apocalyptic counter-knowledge on climate change vis-à-vis geopolitics is generated, validated, and reinforced. In the light of the addressed interdependencies between offline histories and various beliefs narrated within online spaces, it is an important objective of this research to identify the underlying epistemic, cultural, and (geo)political dispositions that are embedded in contemporary apocalyptic conspiracist discourses. To do so, we conducted a critical discourse analysis on textual materials published on American apocalyptic conspiracist websites. While our focus here will be on American evangelical End-Times websites and blogs, we also refer to non-religious sources. Based on the theoretical framework of this book developed above, we argue that all forms of knowledge must be explored as a result of intricate cultural histories that exceed the analytical categories of the religious and secular. We also therefore discuss sources without a clear evangelical apocalyptic origin to illustrate that certain ideas and ways of knowing about the world are not exclusive to American evangelical apocalypticism but rather emerge out of shared traditions and beliefs of the American political right and anti-globalist movements.

Similar to other qualitative research on digital conspiracist End-Times discourses or climate change conspiracism (O'Donnell, 2020a; Robertson, 2018; Sharman, 2014; Dittmer, 2010), we explore a cross-section of recently published material (2015–22) that allowed us to examine how writers construct their respective social reality. In 2015, the issue of climate change became highly relevant within conspiracist spaces because the Paris Climate Agreement, a perceived proof for the imminent one-world government, was adopted this year. The emphasis here is on textual material that is intended for public consumption, like a news or blog article, a YouTube talk show, or a podcast. We approach writers that publish public material as "as spokespersons; focal points rather than *de jure* leaders with an official mandate" (Robertson, 2018: 237).

To identify relevant websites, we employ a combination of different internet-based search methods. We make use of a frequently updated list of the most popular "End-Times" and "Prophecy" websites that are ranked "by traffic, social media followers, domain authority & freshness": Feedspot (FeedSpot Blog, 2018–22a; FeedSpot Blog, 2018–22b). Yet, attention needs to be paid to the prevailing beliefs on each website, as FeedSpot also includes non-evangelical websites. Additionally, FeedSpot's ranking algorithm includes variables that cannot be tested from an outside observer. Some highly ranked websites could be affected by paid placements rather than objective quantitative factors. For example, the popularity of some websites appears to be exaggerated given their low social media followership as well as the small number of authors and posts. Hence, we put an additional focus on Facebook and Twitter followership because such quantitative factors of popularity that can be more reliably monitored in these spaces by external persons.

Moreover, we engage with websites and authors that are mentioned in the academic literature on End-Times conspiracism, like Infowars, RaptureReady, or Hal Lindsey's websites (see, e.g., Robertson, 2018, 2015; Knowles, 2013; Dittmer, 2010). Additionally, we identify websites of interest through Google and DuckDuckGo search queries where we searched the internet for a combination of terms (e.g., "climate change" + "Rapture" + "Antichrist") which could indicate evangelical apocalyptic conspiracist thought. Nevertheless, we found that such search engine results closely resembled the lists of the most popular End-Times and prophecy websites/blogs on FeedSpot. Other websites were found "via a snowball method using blog-roll links" (Sharman, 2014: 162). Robertson (2018) writes that different authors or websites within the conspiracist milieu can be approached as a network of small groups that are connected through a shared ideology rather than any centralized institutional structures. The network structures are illustrated by authors who publish their texts on different websites of the digital American End-Times conspiracist milieu. An example of this is the author Michael Snyder, who posts texts on RaptureReady and EndTimeHeadlines as well as the non-evangelical website Infowars.

However, after identifying a list of relevant digital spaces, or websites, we used Google's site search function to collect articles that address climate change in light of conspiracist End-Times beliefs. In this study, we emphasize spaces that provide a large number of versatile and long texts (over 800 words) on climate change, as they can reveal the underlying motives, ideologies, and epistemic strategies of evangelical apocalyptic conspiracists. Google's site search function allowed us to search a website for relevant terms like "climate change," "global warming," "environment," "greenhouse gas," and "NWO" (e.g., endtimeheadlines.com, "climate change"). We examine texts that clearly emphasize climate change but also texts which only touch upon the subjects to explore how they are put into context with other issues. The required affiliation to the American culture is determined either through the website's publishing information/contact details, the origin of the authors, or a clear thematic emphasis on the American culture. Texts are selected for clearly demonstrating an apocalyptic conspiracist worldview, which, in the most basic sense, describes the counter-knowledge that the end of the world is intentionally caused as the result of malicious plots. In that context, all search results, websites, and publications are assessed manually to ensure that only relevant webpages and publications are included in this research project. Following O'Donnell's (2020b: 3) reasoning, "texts were not chosen for the influence of their authors, which is often difficult to measure," but from a broad set of authors to "identify and unpack discursive tendencies in this body of literature without giving undue weight to the idiosyncrasies of authors or—when applicable—situating such idiosyncrasies in their wider discursive contexts" (2020). Not all analyzed texts are represented in this book, as the emphasis is on selected examples of significant quotes that reveal the epistemic and discursive strategies of digital apocalyptic conspiracism. To facilitate the analysis of videos and podcasts published by "intervangelists"

(Bekkering, 2011) (contemporary video preachers who broadcast online), we used Microsoft Word's automatic transcript function.

Conclusion: Do Your Own Research

It is important to understand the construction of alternative knowledge in digital spaces as an increasing number of people acquire their knowledge online while, at the same time, the trust in mainstream media declines in the United States and other countries (Andersen, Shehata, and Andersson, 2021; Jones, 2004). As a seemingly infinite source of easily accessible knowledge, users can research virtually anything, often detached from the traditional power structures and epistemic authorities of the "physical" world. Yet, people do not simply "do research" on the internet but rather "do their *own* research," a slogan where the possessive pronoun indicates "a broader worldview in which research and science can yield multiple and contestable truths" (Carrion, 2018: 317). For example, an Infowars (2021) article on former American presidents who promote Covid-19 vaccinations requests readers not to trust elite opinions: "Do yourself a favor and don't take medical advice from a washed-out globalist politician—do your own research and come to your own conclusions." In short, to "do your *own* research" implicates a distrust in dominant beliefs accompanied by the endorsement of specific counter-knowledge (Carrion, 2018). In this book, and similar to someone who is doing their "*own* research," we will engage with publicly accessible websites and sources of knowledge that provide alternative knowledge-claims concerning climate change. We address the creation, communication, and the justification of apocalyptic conspiracist knowledge-claims and analyze the discursive practices through which authors attempt to convince their audience of a particular truth-claim or claims.

4

American Evangelicalism, the Environment, and Apocalyptic Conspiracism

Evangelicals in the United States have a long history of interpreting increased governance, on the global scale (McAlister, 2018; Kaplan, 2018; Hummel, 2018; Durbin, 2018; O'Donnell, 2020b) or the national scale (Sutton, 2017, 2012; Dochuk, 2011), as threats to American Christian freedom and as indicators of the rise of the Antichrist. While today, in the public sphere, American evangelicals are often depicted as a homogeneous group of white Christian nationalist Trump followers, it needs to be stressed that not all designated or self-identified American evangelicals endorse conservative politics, nationalism, or radical right-wing conspiracist End-Times discourses.

American evangelicalism is more than just religious doctrine. It is a diverse movement characterized by distinct and sometimes contradictory religious, cultural, and political attitudes (Marsden, 1980; Noll, 2002; Sutton, 2017). Consequently, the term "evangelical" "has produced more debate than agreement" (Worthen, 2014: 3), and a variety of terms have been used in academic debates to address the complexities of American evangelicalism. Depending on the historical and social context, the rather conservative American evangelical traditions, which mostly endorse pessimistic apocalyptic beliefs, have been described as fundamentalists, Christian nationalists, neo-evangelicals, neo-charismatics, radical evangelicals, or anti-modernist Protestant evangelicals, to name a few (Marsden, 1980; Sutton, 2017; Gribben, 2011; O'Donnell, 2021). For the subsequent general introduction on American evangelicalism and its relationship to climate change and the natural environment, we will use the general term "evangelicalism" before specifying which specific evangelical dispositions and sentiments are of interest for this research. The following sections describe the basic features of evangelicalism, its relationship to climate change, and the natural environment more broadly to approach the cultural context out of which apocalyptic conspiracist discourses emerge.

Section I : More than Religious Doctrine, Politics, and the Environment

American evangelicalism is a trans-denominational movement within American Protestant Christianity. About 25 percent of the adult population of the United States identifies as evangelical or born-again Christians (Pew Research Center, 2019, 2015). While in recent decades, the popularity of Protestant Christianity as a whole decreased in the United States, the share of Protestants who identify as evangelical/born-again (59 percent of Protestants in the United States) "is at least as high as it was a decade ago" (Pew Research Center, 2019: 22). About three-quarters of the self-identified evangelicals are white Americans (Pew Research Center, 2015).

In a religious sense, evangelicalism is often approached by using the "Bebbington Quadrilateral" (Steinmetz-Jenkins, 2020: 29), a characterization of evangelicalism coined by the historian David W. Bebbington (1989: 3), who defined four characteristics which separate evangelicalism from other forms of Christianity and mainline Protestantism:

(i) *Conversionism*, the belief that lives need to be changed or transformed when someone finds his or her belief in Jesus, is often referred to as the born-again experience.

(ii) *Activism*, the expression of the gospel in missionary work.

(iii) *Biblicism*, a particular regard for the Bible as the highest authority.

(iv) *Crucicentrism*, the focus on the sacrifice of Jesus Christ on the cross.

This definition of evangelicalism is used by the National Association of Evangelicals (NAE), which comprises more than 45,000 churches from almost 40 different denominations in the United States. The NAE further defines evangelicals as people who "take the Bible seriously and believe in Jesus Christ as Savior and Lord" (NAE, 2017a). The American historian Matthew Sutton (2017: preface, x) describes American evangelicals as Christians who emphasize "the centrality of the Bible, the death and resurrection of Jesus, the necessity of individual conversion, and spreading the faith through missions." Historian Molly Worthen (2014) explains that the authority of the Bible is crucial to evangelicalism. Since American evangelicalism holds that the Bible alone contains the will of God (an idea called *sola scriptura*), an evangelical's faith and practice are guided by Christian scriptures. Nonetheless, it is important for this academic engagement with evangelicalism to note that the interpretation of the Bible differs among denominations, traditions, and movements—so what is understood as the "right" doctrine, or "true" Bible knowledge, is largely subjective and depends on social environments, cultural histories, and societal power relations. Importantly, this centrality on Scriptures has increasingly been challenged by the ingress of neo-charismatic movements (O'Donnell, 2021). Versatile biblical truth-claims, conceptions of God's word, and varying eschatological belief systems within American evangelicalism have implications for cultural, political, and environmentally relevant attitudes.

However, the label "evangelical" is commonly related to much more than the Bebbington Quadrilateral or Protestant doctrine, which emphasizes biblicism. Addressing the evangelical support of Trump's 2016 presidential campaign, the American historian Thomas S. Kidd (2016) observes a "watering-down and politicization" of the word "evangelical" because it became increasingly used by Christian Trump supporters who self-identify as evangelical based on their political attitudes. In an analysis article for the *Washington Post*, Kidd (2016) argues that "evangelical now basically means whites who consider themselves religious and who vote Republican."

This is not a new phenomenon of the Trump era. Marsden (1980, 1991) also observed that historically, the label evangelical has been utilized to describe politically conservative Protestant Christians regardless of religious doctrine. In that context, Margolis (2020) discusses a possible distinction between traditional/believing evangelicals (those who hold specific religious beliefs associated with evangelicalism) and nominal/cultural evangelicals (those who do not hold specific religious beliefs associated with evangelicalism). Still, Margolis (2020: 111) eventually states that even if such a distinction is employed, empirical data still show that both groups were supportive of Donald Trump, and therefore self-identifying as evangelical "does not produce an overestimation of Republican support among white evangelicals." Consequently, Margolis (2020) argues that the term "evangelical" has not become a synonym for conservative politics, but that politics is a part of white evangelicalism, traditionally, but not exclusively, in connection to the Republican Party and conservative politics. In other words, there is no distinction between religious evangelicalism and political evangelicalism, but rather, evangelicalism *is* political, although in a variety of ways (Sturm and Dittmer, 2010).

Such conceptions of evangelicalism support Taylor's (2007) or Asad's (2003, 1993) conceptualizations of religion and the secular. They argue that religion cannot be separated from other dimensions of society while what is perceived as secular is affected by century-old religious traditions (see also Chapter 7 on the interdependencies of religious and secular apocalypticism). Influenced by Bourdieu's work on religious practice and the reciprocal relationship between different fields of society, Asad (1993: 129) writes that religious practice is to be "explained as products of historically distinctive disciplines and forces," including politics, science, economics, and more. Asad (1993) rejects a clear distinction between religion and politics, which he argues is a predominantly Western distinction. In this book and through our engagement with evangelicalism, we follow Asad's conception of religion and argue that American evangelicalism is characterized by the inseparability of religious and non-religious social forces, so that the rather faith-based "Bebbington Quadrilateral" alone cannot hold as a suitable and timely definition of evangelicalism. Historically, the Christian tradition, including American evangelicalism, was always shaped in varying degrees by a wide range of non-religious developments, while, conversely, religion affects the perception of non-religious issues (Fitzgerald, 2011; Martin, 2017). Sutton (2017), Miller (2014), Gribben (2011), and Marsden (1980) show in their in-depth engagements with the history of American evangelicalism that the interpretations of the Scriptures were constantly changing and adapting to developments in the non-religious social dimensions, for instance, national politics, geopolitics, science, and eventually, climate change. Evangelical religion has always responded to contemporary developments from other social forces, as "social change breeds religious change" (Rey, 2014: 75–6), and consequently, secular culture and politics were always a part of evangelical religion.

Due to the versatile influences which shape Christian religious movements and their respective beliefs, American evangelicalism as a whole is more heterogeneous than the often-cited affiliation between evangelicals, conservative politics, and the Republican Party would suggest. Many American evangelicals are not aware of the diversity of their movement (Worthen, 2014) and perceive American evangelicalism as "one-dimensional" (Miller, 2014: 4), even though evangelicalism should be approached as "culturally and politically ambivalent" (Swartz, 2012: 1). Evangelicalism is "a populist movement that transcends denominational lines and ethnicities" (Steinmetz-Jenkins, 2020: 31). An important historical reason for the diversity of American evangelicalism is the lack of hierarchical systems of governance. A "crisis of authority" (Worthen, 2014: 2) caused a fragmentation of American Protestantism (Verter, 2003), in which a wide range of actors tried to govern the discourse on religious and other issues. Consequently, attention needs to be paid to the diversity of American evangelicalism when engaging with its complex religious communities, especially regarding spatial, demographical, and racial variations of this movement within the United States (Wong, 2018a, 2018b, 2015; Carroll, 2012; Dayton and Johnston, 1991).

Race, in particular, is often discussed as an important factor in the political attitudes of American evangelicals. Even though popular news outlets often create headlines which suggest that evangelicalism as a whole considers Trump to be their president and savior, it must be stressed that this mostly applies only to white evangelicals since, in the 2016 US presidential election, "non-white evangelicals mostly voted for Clinton" (Gorski, 2019: 347). Evangelicalism is often wrongly used as a synonym for the Christian right or Christian nationalism, who "believe that the United States was founded as a white Christian nation, and they fear that 'their' nation is being muddied by non-European immigrants, corrupted by 'secular humanists,' and infiltrated by 'radical Islam'" (Gorski, 2019: 348). Steinmetz-Jenkins (2020: 29) highlights the historically established connection between "white supremacy and evangelical political theology" in which particular white evangelicals "have been drawn to right-wing populism" (2020). Therefore, "white evangelicalism" differs profoundly from non-white evangelicalism in terms of political and societal attitudes (Holder and Josephson, 2020; Wong, 2015). In that context, Wong (2018a: 101) observes an "important association between evangelical identity and a politically conservative agenda" among white American evangelicals, whereas non-white evangelicals hold more diverse political attitudes.

In addition to the relevance of race, Carroll (2012) addresses spatial variations within American evangelicalism where evangelical practices in the American south-east, described as the Bible Belt due to the prevailing Christian conservatism, differ from the more liberal-leaning evangelicalism practiced, for instance in California or Seattle. These spatial variations in American evangelicalism can also be found when looking at sites of evangelical learning. For instance, the Fuller Theological Seminary in Pasadena, California, is known for its more liberal teachings and is a central site from which the new evangelical left emerged during the 1970s (Worthen, 2014; Swartz, 2012). By contrast, the Dallas Theological Seminary in Texas is known as a place of conservative evangelicalism and dispensational eschatology (Sutton, 2017).

The overall influence of the diverse Christian movement on the American culture is "profound" (Miller, 2014: 4), as evangelicalism "resides at the very centre of American history" (2014: 7). Sutton (2017: 377) writes that evangelicals "helped to make and break presidential candidates" and contributed to shaping social and political debates in the United States, even though some evangelical communities consider themselves to be "suppressed" (Miller, 2014: 4) and victims of perceived racial inequalities in American society. Many evangelical groups consider themselves outsiders in the American society, whereas outside observers describe evangelicals as part of the American establishment, usually in affiliation with the Republican Party (Schwadel, 2017; Brint and Abrutyn, 2010). This perceived feeling of oppression of evangelicalism is expressed, for instance, in the belief in a secular, liberal federal government that suppresses religious freedom as well as Christian values and threatens the conception of the United States as a Christian nation (Steinmetz-Jenkins, 2020). In recent years, white evangelicals in particular believe that life is "becoming harder" (Pew Research Center, 2016) for them due to the increasing acceptance of religious and sexual diversity, abortion, racial diversity, or the teaching of evolution in public schools. Furthermore, Wong (2018b: 53) writes that "a majority of white evangelicals say that the American culture and way of life have worsened" since the 1950s. Due to the perceived worsening of conditions in the United States, white evangelicals have been susceptible to the American exceptionalism and revivalist nationalism suggested by the "Make America Great Again" slogan employed by Ronald Reagan in 1980, and most recently by Donald Trump (Hummel, 2018).

However, in addition to the politically and religiously conservative white evangelicalism, liberal ideologies became increasingly visible in white and non-white American evangelicalism in the 1960s, forming an "evangelical left." Especially young American evangelicals reanimated liberal trends within American evangelicalism by combining "the rhetoric and tactics of the secular New Left" (Worthen, 2014: 180) with Christian traditions. Social concern and the general good became increasingly important to young evangelicals, in contrast to the traditional evangelical emphasis on the salvation of the individual soul, which focuses on the personal relationship with God (Miller, 2014; Bauman, 2014). The new left-leaning tendencies within American evangelicalism were enforced by the American civil rights movement of the 1960s, to which the young evangelical left contributed. Since political demands in relation to the rights of women and African Americans were crucial to this left-leaning evangelical movement, which emerged at evangelical colleges and state universities, liberal evangelical campaigns were accused by conservative evangelicals of being a political rather than a Christian religious movement. According to Swartz (2012: 3), left-leaning evangelicals "comprise more than a third of American evangelicals and point to the political, theological, and cultural diversity of the movement."

One of the goals of the re-emerged evangelical left was to "be open to the work of God in unexpected places" (Miller, 2014: 148)—for instance, in the natural environment. Many white evangelical Christians campaigned for nuclear disarmament and environmental protection and have justified their position on the grounds that such issues were important in the present age, despite the world falling under God's judgment (Dochuk, 2019). Nonetheless, the political influence of the evangelical left cannot be compared with the significant influence of white conservative or Christian nationalist evangelicalism (Wong, 2018b; Swartz, 2012, 2011). Through their prism of victimization, conservative evangelicals hold and exercise significant cultural and political power, attempting to make the world in the image of their "orthotaxies"—orthodox taxonomies, or right orders, that rule the meaning of the world—that promote "patterns of misogyny, homophobia, Islamophobia, antiblackness, and (settler) colonialism" (O'Donnell, 2020b: 142). Through their own ordered cosmogony and apocalyptic expectations, they foreclose the futures of already precarious lives and, as explored next, environmental action.

Christianity, the Environment, and Evangelicalism

The reciprocal relationship between Christian religion and the environmentally relevant fields of society, may it be climate change, the exploitation of natural resources, or environmental degradation, illustrates the above-mentioned interdependencies between religious and non-religious fields of society. While preexisting religious dispositions can affect the way Christians perceive climate change or think about environmentally relevant practices more generally (Hulme, 2015a, 2015b; Gerten and Bergmann, 2012), the environmental movement also has a profound impact on Christianity and central biblical knowledge-claims. Gottlieb (2004: 584) calls the relationship between religion and climate change/environmentalism a "double-movement": religious actors, communities, and their spiritual values enter the public discourse on climate change and environmental politics while conversely, religious beliefs and practices adapt to climate change and environmentalism. Climate change "challenges and changes images of God and the sacred and their corresponding sociocultural practices" (Bergmann, 2009: 98). Due to the presence of climate change and other environmentally relevant discourses in the public sphere, the Bible, as a central source of knowledge for Christians, and evangelicals in particular, had to provide knowledge-claims on humanity's relationship to the natural environment (Cobb Jr., 2004; Peterson, 2000). In that context, Livingstone (2002b: 353) observes a "Greening of Theology" within the last quarter of the twentieth century and further explains that Christian environmentalism is "at least in part, the retrieval of . . . historical voices" (2002b). Important for the academic, theological, and public debate on the role of the natural environment in Christianity was Lynn White's (1967: 1967) famous essay "The Historical Roots of our Ecological Crisis" in *Science* magazine, which describes the Judeo-Christian tradition and especially Christianity in its Western form as "the most anthropocentric religion the world has seen" (White, 1967: 1205). White (1967) identifies a human-nature dualism in Christianity that "insisted that it is God's will that man exploit nature for his proper ends." The "most influential" (Peterson, 2000: 155) verse from the Bible that is interpreted as supporting human-nature dualism and the belief that God created Earth solely as a resource for humanity is Gen. 1:28:

"And God blessed them, and God said unto them, Be fruitful, and multiply, and replenish the earth, and subdue it: and have dominion over the fish of the sea, and over the fowl of the air, and over every living thing that moveth upon the earth." Even though White neglects the historic as well as spatial variations of Protestant and Catholic movements (Cobb Jr., 2004), his famous essay has initiated a public dialogue and interdisciplinary academic discourse on the relationship between Christianity and nature (Jenkins, 2009; Sideris, 2006; Livingstone, 2002b). The greening of theology and the re-evaluation of Christian traditions, beliefs, and practices illustrate that what is considered biblical truth can change due to influence from non-religious fields of society, as in this case, climate science and the environmental movement.

In the early 1970s, environmentalism entered the field of American evangelicalism as "an increasing number of evangelicals began to emphasize that the right care of the environment was an important Christian concern" (Roberts, 2012: 229). Lowry (2016) and McCammack (2007) address the "greening of theology" in American evangelicalism and identify "eco-evangelicals" (Lowry, 2016: 57) or "green evangelicalism" (Amos, Spears, and Pentina, 2016: 224) within the diverse movement. Much of the research focuses on contrasting interpretations of the Bible, which can also be used to argue for a *Creation Care* approach in which it is a Christian's duty to take care of and be stewards of God's creation. This "stewardship model is framed as a form of direct service to God" (Lowry, 2016: 61, following Jenkins, 2009). The Christian concern for the poor also affects the environmentally relevant beliefs and practices of American evangelicals, as it is argued that climate change and environmental degradation would negatively affect the precarious groups of the world (Kearns, 2014; McCammack, 2007). As such, religiously motivated climate justice also entered the domain of evangelicalism. Evangelical environmentalists are likely to accept "basic global warming science" (Harrington, 2009: 18) to support their concerns about the state of God's creation.

In light of distinct dispositions in American evangelicalism, it is no surprise that "evangelical political elites are deeply divided about how their faith should inform their environmental and political views" (Danielsen, 2013: 199). While several organizations, like the *Evangelical Environmental Network* (EEN) and the NAE, accept the general scientific consensus on anthropogenic climate change and argue for a reduction in greenhouse gas emissions (Roberts, 2012; McCammack, 2007), other evangelical groups reject the idea that humans have the power to alter God's creation and consequently refuse any form of climate protection policies. In that context, Roberts (2012: 227) writes that "perhaps there is no issue in which evangelicals are more divided than over the environment." McKeown (2006, cited in Roberts, 2012) suggested a distinction between "Brown" and "Green" evangelicals, whereas

McCammack (2007) distinguished between "liberal" and "conservative" evangelicals concerning their environmental attitudes. Still, Roberts (2012) notes that a distinction between two separate camps is too simple in light of the diversity of American evangelicalism. Zaleha and Szasz (2015) further suggest that it is more appropriate to look beyond possible denominational labels and terms like "evangelicalism" but instead think of three different categories of Christians (theologically conservative, moderate, and liberal), which also includes non-evangelical Protestants and Catholics.

However, many American evangelicals resist a so-called "greening" of their theology without employing apocalyptic or conspiracist discourses, as not all "skeptics" advance conspiracist knowledge-claims. A growing body of literature engages with skepticism concerning anthropogenic climate change among American evangelicals (Zaleha and Szasz, 2015, 2014; Kearns, 2014, 2012; Roberts, 2012; Veldman, 2019; Kowalewski, 2023). Within evangelical anthropogenic climate change skepticism (ACC-skepticism), it is argued in the tradition of Christian anthropocentrism that the book of Genesis should be understood as a divine command to subdue the natural environment because Earth was created by God to serve human needs. To some, acknowledgment of environmental decline and resulting reform is an act of liberal Christianity that betrays the gospel (Northcott, 2004: 60). Moreover, the scientific consensus on anthropogenic climate change is rejected. In particular, it is argued that the "post-normal" climate science, often described as "junk science" by evangelical ACC-skeptics, produces inaccurate results and is politically motivated (Roberts, 2012). Evangelical ACC-skeptic organizations like the Cornwall Alliance, which is frequently the academic focus of studies on evangelical ACC-skepticism, hold that climate science, like all forms of science, should follow the scientific ideals and empirical philosophy of Francis Bacon or similar epistemic strategies (e.g., Thomas Reid's *Scottish Common Sense Realism*). The Baconian method is an inductive approach to science in which observations in the real world are used to make generalizations, while the generation of knowledge-claims which exceed observable facts should be avoided. The computer-based modeling of future climates and predictive climate science as practiced by the IPCC contradicts this epistemological perspective based on real-life observations. Generally, American evangelicals do not reject science per se but attempt to unveil the complex laws God has implemented into nature by using certain scientific philosophies, like the Baconian method, and further argue that science can confirm biblical truths (Evans, 2011; Roberts, 2012; Noll, 2002). Moreover, evangelicals are more likely to reject scientific findings when they implicate restrictive political measures (Kearns, 2012). Likewise, Noll (2002) explains that evangelical criticism of modern science is often a resistance to social change rather than a general rejection of scientific methods. The field of politics and the socioeconomic framing of climate change are often described as "the real battleground" (Kearns, 2014: 167) of American evangelical climate change debates. The climate change debate frequently acts as a "proxy" (Turnpenny, 2012: 403) for political and socioeconomic debates, and ACC-skepticism is frequently accompanied by the advocacy of free-market systems, laissez-faire politics, and decreased regulation.

As noted before, academic literature discusses terms like "conservative" or "liberal," "green" or "brown" to capture different environmental attitudes among American evangelicals. However, our engagement with textual sources from the Cornwall Alliance and the EEN shows that even white conservative, Republican-leaning, anti-regulation evangelicals who emphasize the authority of the Bible can be "green" as well. In the EEN's book *Caring for Creation: The Evangelical's Guide to Climate Change and a Healthy Environment*, the authors Mitch Hescox and Paul Douglas (2016: 17, italics in original) identify themselves as Republicans and political conservatives who are looking for a "pragmatic, commonsense, *conservative* response to climate change." The authors explain that "caring for God's creation is a biblical imperative—it's about living rightly with God and each other" (2016: 147). The EEN's authors still defend an anthropocentric worldview, as Hescox and Douglas (2016: 18) state, for example, that "climate change is not about polar bears—it's about our kids." Human-nature dualism and anthropocentrism in the EEN's evangelical environmentalism are also observed by Lowry (2016). Furthermore, the EEN avoids using the IPCC's predictive climate science as a source of knowledge and argues instead that science cannot show "what comes next" following a scientific philosophy reminiscent of the Baconian ideal (Hescox and Douglas, 2016: 90). The EEN provides a figure based on past observations, as it shows that rising CO_2 levels correlate with an increase in global temperature between 1980 and 2010. The observed correlation between CO_2 and global warming allows the EEN to argue in favor of anthropogenic climate change without referring to the "post-normal" IPCC science.

Although the EEN supports the Paris Climate Agreement (Hescox and Ball, 2017), the evangelical organization rejects profound governmental interventions as the authors "are searching for a way to overcome climate change in harmony with our conservative values to create a better America" (Hescox and Ball, 2017: 101). The EEN assumes that a free-market system will eventually regulate itself in favor of the climate and environment. Nonetheless, Lowry (2016) argues that this anthropocentric, politically, and theologically conservative approach of the EEN is likely used to appeal to a conservative American evangelical audience. An important message of the EEN and similar organizations is that Creation Care can be consistent with conservative evangelical beliefs like biblicism, anthropocentrism, the advocacy of free-market politics, and trust in epistemic strategies which use the Bible and are in accordance with inductive scientific methods.

While there are many other entanglements of evangelicalism, politics, and climate change worthy of discussion, the purpose of this section was to show that even without any notion of the Antichrist, the apocalypse, or conspiracist knowledge-claims, the field of evangelical environmental attitudes is versatile and characterized by distinct interpretations of the Bible, concurring political attitudes, and discussions on valid scientific philosophies. But to examine discourses concerning the Antichrist's supposed climate change deception, it is necessary to engage with an important feature of American evangelicalism, which might be "the most ecologically decisive component of a theological system" (Curry-Roper, 1990: 159)—Christian eschatology.

Evangelical Eschatology and the Environment: Postmillennialism, Premillennialism, and Dispensationalism

The term "eschatology" describes discourses about "last things, the furthest imaginable extensions of human and cosmic destiny" (O'Leary, 1998: 5) and addresses Christian imaginations of the future. Gribben (2011: 11) states that traditionally, eschatology means discourses about the "four last things— death, judgment, heaven and hell"—but that the popular usage of the term now refers to end-of-the-world beliefs more broadly. As the "theology of the last things" (Boyer, 1992: 3), many evangelical Christians place eschatology at the center of their belief system and search the Bible for clues to future events. Addressing eschatology in evangelicalism, Gribben (2011: 11) writes that "evangelical eschatology can be either pessimistic, in its expectation of the apocalypse, or optimistic, in its expectation of a golden age," the *millennium*. In the most basic sense, there are "two revisions of millennialism" (Marsden, 1980: 48) in American evangelicalism that ask if Jesus Christ is going to return before (pre) the millennium or after (post) the millennium. These different truth-claims regarding the end of times and Jesus' return to Earth in American evangelicalism are based on concurring Bible interpretations, which, like other elements in American evangelicalism, stand in a reciprocal relationship to other aspects of society. Evangelical eschatological knowledge-claims change over time and respond to historical developments while, conversely, eschatological dispositions affect how adherents perceive the world around them (Gallagher, 2012; Stunt, 2012; Dittmer and Sturm, 2010; Sturm, 2012; Boyer, 1992). The common distinction between "pre" and "post" millennialism is not mutually exclusive, and overlap between these eschatological ideas exists. Consequently, it is important to consider the historical context of evangelical eschatology as key beliefs of the same millennial system develop and change over time (Sutton, 2017; Gribben, 2011; Curry-Roper, 1990).

The above-mentioned "optimistic" eschatological belief system in American evangelicalism is *postmillennialism*. Postmillennialism holds that Jesus is going to return after (*post*) the millennium. Adherents traditionally advocate that they can "help inaugurate [the millennium] through their own good works" (Sutton, 2017: 14). It is a popular postmillennial belief that humanity has fallen into sin in the past and is now responsible for defeating evil forces and establishing God's kingdom on Earth through their own behavior. Through Christian teachings, the spread of the gospel, and sometimes also cultural and scientific progress, postmillennial evangelicals attempt to precipitate the millennium on Earth and thereby prepare the world for Jesus' return (Worthen, 2014; Curry-Roper, 1990). The postmillennial belief of God's kingdom on Earth has important implications for environmental behavior. Here, Curry-Roper (1990: 167) argues that the aspired establishment of God's kingdom and the restoration of Earth after humanity's past fall into sin likely result in environmentally friendly behavior as "history is seen as progressive, with this very earth as the present and future home of mankind." The previously mentioned EEN (Hescox and Douglas, 2016: 75) is a good example for postmillennial environmentalism as the authors stress that the natural environment needs to be protected and healed to establish God's kingdom on this Earth:

> The whole of creation is to be healed through and by Jesus the Christ. It's not escaping the earth into heaven but heaven coming to earth and the return to the beauty of the created order: the old becoming new, being the fulfilment of the kingdom. Jesus calls us to restore his will—his kingdom—on earth as it is in heaven.

Nevertheless, there is no simple causal relationship which postulates that advocates of postmillennial eschatology automatically support climate protection and environmentalism. Amos, Spears, and Pentina (2016: 246) show that a postmillennial eschatology can result in a rejection of environmental and climate protection policies because "environmentalism erodes social well-being, prohibiting the progression toward a Golden Age of prosperity," which is the postmillennial idea of God's kingdom on Earth. It is argued that the higher cost of renewable energies or any form of environmental or carbon taxation would adversely affect the nation's and world's precarious populations, and therefore environmentalism hinders the establishment of prosperity. However, the EEN further argue that American evangelicals should "stop worrying about who gets into heaven and start building the kingdom of God on earth" (2016: 78), which, like the previous quote, is a clear hint at the other dominant eschatological belief system that is frequently associated with rather environmentalist damaging attitudes: dispensational premillennialism.

Premillennialism teaches that Jesus' Second Coming is going to take place before (*pre*) the millennium. Here, *dispensational premillennialism* or *dispensationalism*, the most dominant eschatological system in contemporary American evangelicalism, argues that "the world is getting worse and uses environmental problems to predict its end" (Curry-Roper, 1990: 166). Curry-Roper (1990) explains that adherents of dispensationalism sometimes even welcome environmental degradation and natural disasters because these developments are interpreted as indicating the imminent Rapture, an event in which true believers rise into Heaven to meet Christ. Still, such an appreciation of environmental change to the worse is criticized by the EEN. However, many dispensationalists strive for their Rapture into Heaven while it is believed that the current Earth will be destroyed eventually (Gribben, 2024). Consequently, these dispensationalists see no need for environmental action or climate protection (Sturm and Lustig, 2022). In addition, it is argued in the tradition of evangelical ACC-skepticism that humans do not have the power to alter God's creation, as only he can destroy it during the End-Times. Zaleha and Szasz (2015: 25) also argue that dispensational premillennialism forms "part of the bedrock of conservative Christian anti-environmentalism." Similarly, Hulme (2009: 155) explains that a dispensational premillennial mindset is "less likely to stimulate the desire and behaviour to avoid such an outcome," which is the End-Times accompanied by environmental degradation.

Certainly, the expectation of the imminent destruction of the contemporary Earth due to the biblical End-Times is a relevant component of evangelical premillennial ACC-skepticism. However, we argue that another element of dispensational eschatology, the belief in the Antichrist's global reign during a period called the Tribulation—"a final seven-year period of terrible suffering during which the Antichrist persecutes believers and God pours judgement on the world" (Gribben, 2011: xiv)—is the most important driver for the rejection of mainstream climate science, environmentalism, and global climate protection policies. To remind us of Asad's (1993) Bourdieuan notion of religion, which sees religion as influenced by society, it is important to engage with dispensationalism not just as a religious doctrine but as a stance which reflects political, social, and scientific beliefs. The most popular American evangelical eschatological belief system in the last century emphasizes the "interrelationships between secular politics, divine history and the end times" (Sutton, 2017: 16). In that context, Hummel (2020: 288) identifies "three levels of the apocalypse in American evangelicalism—theology, culture, and politics" as premillennialism functions "as an engine for theological debate, cultural production, and political activism." In short, dispensationalism provides not just a religious End-Time doctrine about the physical decay of the planet or individual salvation but also prophecies of cultural and political developments as a "social decay" (Worthen, 2014: 178), which indicate the imminent biblical apocalypse. Dispensationalism holds that several signs are to "precede the return of Christ, including the rapture of believers [. . .] wars, famines, earthquakes, the appearance of the antichrist, and the great tribulation" (Curry-Roper, 1990: 159). However, the belief that signs would precede the Rapture only became popular within the dispensationalist imagination after the publication of Hal Lindsey's best-selling book, *The Late Great Planet Earth* (1970), which emphasized the establishment of Israel in 1948 as the most important sign of the times since the resurrection of Christ. As Gribben (2009: 90) makes clear concerning this reformed emphasis on pre-Rapture sign identification in dispensationalism, "Israel and atomic weaponry were proof the end was near."

Political developments have popularized and seemingly legitimized dispensational eschatology. Marsden (1980) describes dispensationalism as an intellectual product of Christian fundamentalism of the nineteenth century that became increasingly popular after the American Civil War, an event which seemed to confirm social decay. In contrast, Sutton (2017) sees the World Wars and other geopolitical events of the early twentieth century, in particular those which led to the establishment of the Israeli state, as popularizers of the dispensational worldview as it expects global warfare and the return of Jews to the Holy Land. Sutton (2017), Gribben (2024), Marsden (1980) trace the popularization of dispensationalism in the nineteenth century primarily to the person of John Nelson Darby.

Mostly based on interpretations of the book of Daniel, Darby divided human history into seven distinct dispensations, which will unfold chronologically to precede the Millennial Kingdom. Thereby, Darby constructed a millennial "timetable" (Gribben, 2011: 84) and propagated that humanity currently lives in the Church Age, the sixth and penultimate dispensation in which humanity will reject Christ, and therefore, peace and prosperity cannot be established. Today, this expected rejection of Christ is frequently applied to the environmental movement, which is perceived as un-Christian worship of the Earth. Following the downfall of humanity in the Church Age, the Great Tribulation, a seven-year period of the Antichrist's global reign, will occur. This dispensational expectation of a global reign of the Antichrist is an important foundation for the rejection of global governance and international institutions. At the beginning of the Tribulation period, according to Darby, the Rapture of true believers into Heaven will occur to "meet Christ in the air" (Marsden, 1980: 52). Nonetheless, the Great Tribulation is not a dispensation in itself but a period of time between the Church Age and the final dispensation of the Millennial Kingdom. The Millennial Kingdom will be initiated by Jesus' Second Coming at the end of the 7-year Tribulation and his defeat of the Antichrist.

After the Millennium, Satan will finally be defeated. The diversity of American evangelicalism also applies to dispensationalism. Not all dispensationalists, then and today, follow Darby's teachings, as "dispensationalists have never reached total consensus on some of the elements of their theology" (Sutton, 2017: 17). Disagreement exists, for instance, concerning the time of the Rapture. Some dispensationalists think that they will get raptured at the beginning of the seven-year Tribulation, whereas other ("post-tribulationists") believe that the Rapture will occur mid-tribulation after three and a half years, and some think that all Christians have to suffer the full seven years of the Tribulation and will get raptured when Jesus defeats the Antichrist.

Dispensational premillennialism is now largely synonymous with premillennialism due to its popularity among American evangelicals, and henceforth, we will refer to dispensationalism as "premillennialism." Moreover, premillennialism is usually understood as evangelical apocalypticism, as a pessimistic and deterministic discourse that propagates the sudden end of the contemporary world and the revelation of human destiny in a rather pessimistic sense. However, it should be made clear that such eschatological categorizations have become, to use Brubaker's terminology, more categories of analysis than categories of practice. That is to say, the boundaries between the "pre" and "post" millennialisms are not as stark as commonly thought, and as we explore below, these eschatologies are open to diverse "secular" and theological influences and incorporations. Nevertheless, and for this book, evangelical apocalypticism is at core premillennial eschatology which, in short, expects the imminent biblical End-Times, the rise of the Antichrist, and the Return of Jesus before the millennium. Out of the diverse field of American evangelicalism, contemporary premillennialism or evangelical apocalypticism constitute the focus of the subsequent chapters. Rather than engaging with "Christian fundamentalists," "conservative," or "brown" evangelicals, or any of the other terms mentioned in the previous sections, this study engages with American evangelical apocalypticism, or more precisely, what Sutton (2017: 3) calls "radical apocalyptic evangelicalism": those American evangelicals who put premillennial eschatology at the center of their belief and perceive the supposedly degrading world around them in the light of the always imminent biblical End-Times.

Section II: Evangelical Apocalyptic Conspiracism

The previous sections sketched out the wider field of American evangelicalism, its diversity, environmental attitudes, and the basic characteristics of eschatological belief systems to establish a wider understanding of the background out of which evangelical apocalyptic conspiracist discourses on climate change emerge. The field of interest for this study was narrowed down from the versatile field of American evangelicalisms to contemporary evangelical apocalypticism, but a further specification is required since not all premillennial evangelicals are the focus of this study, but only those who postulate a conspiracist worldview in addition to their religious apocalypticism. It is important to conceptualize evangelical conspiracist End-Times thought as knowledge or knowledge-entailing discourses. Apocalyptic conspiracism is the discursive field in which apocalyptic and conspiracist discourses are merged into one coherent belief system that propagates counter-hegemonic knowledge-claims about a profound and imminent transformation of the physical and social environment, intentionally precipitated by evil, all-powerful cabals. The term "apocalyptic" captures two distinct fundamental features of apocalyptic conspiracism. First, apocalyptic discourses prophesy imminent catastrophe and the end of the contemporary world, accompanied by the abolition of established power structures. This popular understanding of apocalypse emphasizes the destruction of Earth and humanity, impending disaster and suffering, usually initiated by supernatural powers, although the previously discussed examples of eco-apocalypticism or imaginations of a nuclear apocalypse show that humanity can also precipitate the End-Times. Second, apocalypticism is also a "mode of knowing" (Stewart and Harding, 1999: 286) that reveals hidden knowledge about the future, human destiny, and an ongoing conflict between good and evil (DiTommaso, 2020; Dittmer and Sturm, 2010; O'Leary, 1998).

Conspiracism is the belief that human destiny is controlled by a hidden group of powerful, evil conspirators who design complex, multi-layered plots to govern the course of human history for their respective benefits (Barkun, 2013; Wilson, 2017). In contrast to an individual conspiracy theory, conspiracism means a worldview that assumes occult powers operate behind the scenes of history. Thereby, "the apocalypse lends itself well to conspiratorial thinking" (Sturm and Albrecht, 2021a: 123), as both discourses, apocalypticism and conspiracism, "satisfy the need for an all-embracing interpretation of history and create meaning in an increasingly secular, disenchanted world" (Hagemeister, 2018: 424).

Field of Investigation

Apocalyptic conspiracist discourses exist in religious and non-religious variations, but for the following chapters, the emphasis is on American evangelical apocalyptic conspiracism. Here, it is important to keep in mind that evangelical premillennialism and non-religious (apocalyptic) conspiracism stand in a reciprocal relation to each other. The American evangelical premillennial tradition provides a religious framework which is sustainable and flexible enough to reflect any crisis and embed contemporary conspiracist knowledge-claims, while, conversely, non-religious conspiracist discourses borrow from religious millennialism (Sturm, 2006; Howard, 2006; Barkun, 2013; Robertson, 2016). Wilson (2017: 413) explains that the religious and political potential within evangelical apocalyptic thinking "amplify the potency of each position."

With regard to adherents of apocalyptic conspiracist interpretations, Ed Stetzer's (2017), now much-forwarded paper in *Christianity Today*, lambasted evangelicals for their uncritical support of Trump and his "alternative facts." Stetzer also argues that evangelicals have a predisposition to conspiracy theories, an observation the authors think requires further analytical development (cf. Keeley, 2007). First, and as Stetzer suggests, American evangelicals, especially those who proclaim an uncritical literal reading of the Bible, are more likely to believe in conspiracy theories. Their belief in being elect or having special revelation endows a belief in the exceptional knowledge that is understood only by chosen people, which is inherent in conspiracism. Second, the deep partisanism of American politics, where overwhelmingly evangelicals support the Republican Party, has become a form of religious nationalism for the "sacred will of the Republican Party" (Sturm, 2018: 14), of which 38 percent of Republican voters are evangelicals (Pew, 2015). The Republican leader, President Trump, often hints at the validity of conspiracies by promoting them or inventing them, from inauguration crowd sizes to questioning the deadliness of Covid-19 and curing it by ingesting disinfectant and, of course, the veracity of climate science (Uscinski, 2020: 111–19). And third, many evangelicals believe that evil is ubiquitous and always plotting in grand, orchestrated ways toward a teleological finale.

In the following, we discuss four key characteristics of the American evangelical End-Times conspiracism of the political right: first, the imagination of the apocalyptic evil; second, the expected transformation; third, the assumption of a superconspiracy; and fourth, the counter-hegemonic resistance to the mainstream.

The Apocalyptic Evil

Contrasting non-religious apocalyptic conspiracist speculations about the true source of the world's evil, the American premillennial tradition provides a certain imagery of apocalyptic evil: the Antichrist, Satan's representative on Earth. As an established belief within the American evangelical premillennial movement, it is an unquestionable truth for adherents that the Antichrist will rise to global power before Jesus returns to Earth to defeat him. Here, it was in particular Darby's dispensational eschatology which pushed the Antichrist to the center of American evangelical millennialism (Hummel, 2020; Barkun, 2013). Nevertheless, the *Feindbild* of the Antichrist is flexible and can be projected onto different personalities, nations, and systems of control, as premillennial evangelicals are "continuously refashioning a new enemy" (Sturm, 2010: 150). The imagination of a larger network, or system, which serves the Antichrist became increasingly popular as it was connected to a variety of individuals or institutions to the apocalyptic evil (Barkun, 2013).

As apocalyptic imaginations usually reflect contemporary societal anxieties and crises (O'Leary, 1998; Boyer, 1992), global communism was believed to be the instrument for the Antichrist's world domination during the Cold War. But in the 1990s, "evangelical concern about global government has shifted from an explicitly anti-Soviet mentality to more general fears about the so-called 'New World Order'" (Wood and Douglas, 2018: 96). Gribben (2004: 86) also states that after the Cold War, premillennialism increasingly drew attention to globalism and the alleged "dangers of a New world Order" as it was assumed that the UN would establish the Antichrist's one-world government. Traditionally, the rule of the Antichrist was mostly applied to the ten-nation kingdom within the boundaries of the Roman Empire, as described in the book of Daniel, but it is now widely assumed that the Antichrist will rise "to helm a one-world government" (Hummel, 2020: 293). Here, the fictional *Left Behind* (1995–2007) novels popularized the idea of an Antichrist-led UN (McAlister, 2003; Gribben, 2004; Frykholm, 2004; Dittmer and Spears, 2009). Still, Boyer (1992: 271) contends that conspiracist knowledge-claims, often addressing supposed anti-Christian globalism, were one of the key themes of evangelical prophecy after the Second World War, eventually ascertaining a "conspiracy thinking endemic in postwar prophecy" since "premillennialism, with its conviction that a single powerful figure would soon rule the world, obviously reinforced such a conspiratorial worldview and drove it to a pitch of high intensity" (1992).

Sutton (2017) argues that apocalyptic imaginations concerning anti-Christian intergovernmental institutions entered the field of American premillennialism in the early twentieth century, specifically with the League of Nations, which was then perceived as a possible candidate to usher in the Antichrist's global reign. Nonetheless, Herman (2001) contradicts Sutton, claiming that the United Nations was perceived as relatively unimportant in American premillennialism until the 1990s. An emphasis was instead put on the European Economic Community (EEC, and later the EU) because it was seen as the successor to the Roman Empire, which is directly named in the Bible. The geographical similarity to the Roman Empire made the supranational institution a "particularly attractive candidate" (Barkun, 2013: 44) for the Antichrist system, as the

Revived Roman Empire, or the ten-nation alliance as described in Dan. 7:23-24.[1] Hal Lindsey's *Late Great Planet Earth* (1970) in particular drew attention to the EEC as the Antichrist's system (Sturm and Albrecht, 2021b).

These distinct imaginations of the Antichrist's institution show that the concept of the apocalyptic evil in evangelical apocalyptic conspiracism is flexible enough to be projected onto different individuals or institutions, reflecting contemporary anxieties and *Feindbilder* (Barkun, 2013, 2010; Howard, 2006). While it is certain that the Antichrist will rise to power, there have always been debates about the identity of the Antichrist. American presidents like Roosevelt (Sutton, 2012) or Obama (Dittmer, 2010), the pope (Herman, 2001; Werly, 1977), or even a supercomputer (Boyer, 1992), were discussed as possible incarnations of the Antichrist. Certain political ideologies which indicate a high level of governmental authority and control, like communism, were interpreted as tools of the Antichrist to acquire power. Previous research on the interdependencies between evangelical apocalypticism and conspiracism shows that the resistance to socialism, communism, the American federal government, global governance, and anti-Catholicism is often driven by imaginations of apocalyptic evil (Sutton, 2017; Herman, 2001, 2000; O'Leary, 1998; Boyer, 1992; Werly, 1977; Hofstadter, 1964).

In that context, conceptions of "right" and "wrong" political or cultural attitudes become tied to the Manichean good-evil dualisms. American nationalism and exceptionalism thereby became incorporated into premillennial eschatology as it is believed that the US and American premillennial evangelical Christians—as "Warriors of Jesus" (Werly, 1977: 42)—fight evil globalists, seemingly false religions (at times including Catholicism), and socialists who are controlled by the Antichrist. Even though evangelical apocalyptic conspiracism mostly propagates the existence of external threats to American sovereignty, it is not a simple, spatial, us-versus-them discourse; America against the rest of the world, as an Evil Other (Sturm, 2010) is sometimes assumed within the United States. This can be the so-called Deep State (O'Donnell, 2020a), a secret apparatus which controls the American federal government in favor of the globalist Antichrist, or American individuals, organizations, and states that are said to be complicit in implementing the anti-Christian world government. Here, it is assumed that some intentionally cooperate with the apocalyptic evil while others fall for the Antichrist's deception and his promises of global peace, a sustainable economy, and equality.

The Apocalyptic Transformation(s)

A pivotal function of all apocalyptic discourses is to prophesy "endings, overturnings, and originary moments," a conversion from the current world order toward the new, post-apocalyptic world (Stewart and Harding, 1999: 286). In evangelical apocalyptic conspiracist discourses, such overturnings are addressed in two distinct, though interrelated, ways. The first apocalyptic transformation entails the superconspiracy that enables an authoritative global dictatorship, in the tradition of non-religious anti-globalist conspiracism, whereas the second specifically addresses the biblical End-Times, including the Rapture and Jesus' Second Coming.

The first apocalypse is the end of the current world order precipitated by anti-Christian forces and the subsequent global dictatorship of the apocalyptic evil, today often designated as the NWO. This apocalypse is the focus of many contemporary evangelical apocalyptic conspiracist discourses. Although the idea of a global dictatorship during the End-Times existed for more than a century in the realm of American premillennialism (Sutton, 2017; Boyer, 1992), it was in particular the radical televangelists and preachers Pat Robertson and John Hagee who, after the end of the Cold War, introduced premillennialism to the NWO concept of the American conspiracist right (Barkun, 2010). Drawing from anti-federal and anti-globalist theories from the American right, evangelical apocalyptic conspiracists argue that satanic forces strive for the dissolution of nation-states through conspiracist plots to establish an authoritarian NWO (Wilson, 2017; Barkun, 2013, 2010). In Christian versions of this discourse, the NWO is led by the Antichrist, whereas non-religious variants assign the role of the evil to the Illuminati or supposed Jewish cabals like the Rothschild family, to name a few examples.

However, the state of the world during the Great Tribulation became equivalent to the one-world government imagined by right-wing conspiracists. Global crises like climate change or Covid-19 are believed to be hoaxes used to justify freedom-restricting measures and the accumulation of power in a single global institution (Sturm and Albrecht, 2021a). Moreover, American evangelical apocalyptic conspiracists imagine a dreadful post-apocalyptic world by borrowing imaginaries from the American right, such as FEMA concentration camps or Black Helicopters that enforce the NWO's dictatorship and persecute non-compliant Americans and Christians (Barkun, 2013; McAlister, 2003; Spark, 2001). The newest addition to evangelical apocalypticism, as we will show later, is the scenario of a *Great Reset*, which originates from QAnon and Alex Jones and describes the abolition of existing power structures and the NWO's takeover of the world's economy during the Covid-19 lockdowns. Nonetheless, conspiracist evangelicals also populated the QAnon movement and contributed to shaping the movement's key beliefs, highlighting a reciprocal relationship between American evangelicalism and right-wing conspiracism (Bond and Neville-Shepard, 2021).

The second transformation, or series of transformations, of evangelical apocalyptic conspiracism addresses the biblical End-Times in the dispensational imagination, which includes the Rapture, the seven-year Tribulation during which the Antichrist rules over Earth, followed by the Return of Jesus to Earth, and eventually the establishment of the Millennial Kingdom and the creation of New Heaven and New Earth. The specific sequence of the events is disputed within different strands of contemporary American premillennialism. In particular, the timing of the Rapture, "a dramatic experience in which all living Christians will mysteriously vanish from the earth and the dead will rise to heaven" (Sutton, 2017: 18), is the subject of discussions, as mentioned. Still, the Rapture is likely the most important penultimate apocalyptic event for premillennial evangelicals, as they believe that all "true Christians," which of course, they believe they are, will meet Christ in Heaven and experience ultimate salvation.

However, it could be argued that the first overturning, the one the superconspiracy aims for, is just the first of many transformations in the biblical End-Times script since contemporary premillennialism portrays the Antichrist's global dictatorship as a requirement for the Rapture, Tribulation, and Jesus' Return to Earth. However, evangelical discourses concerning the NWO, the Great Reset, and climate change conspiracism resemble their non-religious counterparts in the majority of arguments, except for the source of the evil. Due to the non-religious origin of most knowledge-claims concerning the establishment of the NWO, we hold that it is more appropriate to discuss separate apocalyptic transformations in contemporary evangelical apocalyptic conspiracism: first, the establishment of the one-world government through a cultural and political superconspiracy, and second, the biblical End-Times as described in the Bible.

The Superconspiracy

The apocalyptic superconspiracy is the hidden development which will usher in the End-Times. The rise of the Antichrist is inevitable for adherents of premillennialism, but for a long time, the question remained as to exactly how he would rise to power. While some American evangelicals in the first half of the twentieth century assumed that the World Wars could accommodate the rise of the Antichrist in a scenario where war and violence would be used to accumulate global power (Sutton, 2017), it is now believed that the Antichrist will use lies instead of violence to establish a global dictatorship. In contemporary evangelical apocalyptic conspiracism, it is now prophetic knowledge that the Antichrist will establish a global reign through deception, or a series of deceptions, accompanied by the rise of several false prophets (Barkun, 2013). The Bible provides several verses which can be interpreted as prophesying "the deception or seduction by the Evil One(s)" (Hagemeister,

2018: 423). Mentions of deceptions, lies, and false prophets, for instance, in Mt. 24:4-5,[2] 2 Jn 7,[3] Rev. 19:20,[4] or 2 Thess. 2:9[5] announce that an evil deceiver, the Antichrist, will acquire trust, control, and a large following through lies. Barkun (2013: 44) explains:

> The Antichrist folklore consistently emphasized his capacity for deception and control; indeed, it became an unquestioned tenet of dispensationalism that the world would initially welcome the Antichrist as a charismatic peacemaker whose diabolical designs would remain hidden until he had achieved total power.

This apocalyptic expectation of deception is flexible enough to be applied to a wide range of contemporary developments and movements, as "world events are woven into the fabric of biblical prophecy, and 'signs of the End Times' are frequently perceived in the world around us" (Wilson, 2017: 418). Since premillennialism further teaches that humanity will abandon God and Christ during the Church Age (the final dispensation before the Rapture and Tribulation), many adherents expect the rise of a "false" religion to be a result of the apocalyptic deception (Durbin, 2021). Environmentalism, as the supposed anti-Christian worship of Earth based on the testimony of alleged "false prophets" (that being environmental activists and climate scientists), satisfies the premillennial expectation of a false religion established through deceptions and therefore leads Christian believers astray from God.

However, in light of the magnitude of the Antichrist's apocalyptic reign, a single conspiracy would hardly pave the way for global power. Rather, a large, complex web of deceptions and plots is required, a superconspiracy which "incorporate[s] many individual theories into a larger plot" (Wood and Douglas, 2018: 96). The *conspiracism* in apocalyptic conspiracism addresses discourses of a superconspiracy which "attribute all of the world's evil to the activities of a single plot, or set of plots" (Barkun, 2013: 7). This global magnitude and the focus on not some evil individual or group but on *the* global evil makes conspiracism compatible with apocalypticism, as both discursive fields postulate "a hidden, overwhelming power that suffuses history, leaves traces for believers to find, and drives history towards a goal" (Roberston et al., 2019: 3).

The belief in a superconspiracy allows adherents to endorse a variety of different conspiracist beliefs, as individual conspiracist plots are not mutually exclusive, but are understood as a network of different more or less successful attempts to fully establish the global reign of evil. Wilson (2017: 422) explains that the idea of a superconspiracy "has become increasingly pervasive over the past thirty or forty years." As mentioned above, the NWO theory turned into the "common ground for religious and secular conspiracy theorists" (Barkun, 2013: 63). Although evangelical apocalyptic conspiracists do not necessarily employ the term "NWO" or "New World Order," the anti-globalist right-wing discourse of an alleged global superconspiracy, the goal of which is the dissolution of nation-states, drives much of contemporary evangelical resistance to globalism. In the most popular recent NWO discourses on climate change and the Covid-19 pandemic, it is argued that secret conspirators intentionally create or fabricate crises, like climate change or a pandemic, to then create acceptance for restrictive measures which are believed to be the solution for the crises. Global climate change policies, lockdowns, or vaccinations are then perceived as tools of the Antichrist to establish his reign.

The Counter-Hegemonic Resistance and Epistemology

Power-knowledge hierarchies, or more precisely, *perceived* power-knowledge hierarchies, are vital to apocalyptic conspiracism. According to Barkun (2013) and Robertson (2018, 2016), apocalyptic conspiracist discourses employ suppressed knowledge-claims and counter-hegemonic epistemic strategies to attack political and epistemic institutions in power. Still, in light of the previously described power of American evangelicals in American society and politics (Sutton, 2017; Miller, 2014), we argue that evangelicals do not constitute an opposition to dominant knowledge-claims and institutions in power but rather are part of the constituency of power in the United States. In particular, during Trump's presidency, apocalyptic and conspiracist knowledge-claims became increasingly legitimized and powerful as they affected concrete political decisions. For instance, Trump's withdrawal from the Paris Climate Agreement was partially justified through climate change conspiracism. The former American president argued that the "agreement is less about the climate and more about other countries gaining a financial advantage over the United States" (White House, 2017), thereby echoing dominant nationalist and anti-globalist conspiracy theories about climate change. Yet, similar trends can be observed since 9/11 and George W. Bush's presidency.

Robertson et al. (2018: 3) explain that "a religious group may self-consciously and deliberately cast itself as an out-group—whether it is objectively disenfranchised or not—and launch its own counter-epistemic claims against perceived hegemony in order to claim its own space, its own power." This applies to American evangelicalism more broadly and, in particular, to evangelical apocalyptic conspiracism. As mentioned earlier, many American evangelicals feel suppressed in a supposedly increasingly un-Christian and diverse non-white American society (Steinmetz-Jenkins, 2020; Wong, 2018b; Hummel, 2016). Here, evangelical apocalyptic conspiracists perceive themselves as part of a resistance to what an alleged globalist, liberal, and un-Christian mainstream which threatens the United States as an exceptional, white, and Christian nation. As we will show in later chapters, evangelical apocalyptic conspiracists portray themselves as outsiders, victims, and suppressed minorities in American society, what Nietzsche might have likened to "ressentiment" (Connolly, 2008). It is argued that their knowledge-claims as well as their source of knowledge, in particular the Bible, are rejected by the liberal mainstream.

Still, in light of the deterministic characteristics of apocalyptic prophecy (McAlister, 2020), the counter-hegemonic political resistance embedded in premillennial thought seems somewhat paradoxical. Although it is believed that the rise of the Antichrist and the biblical End-Times are inevitable, evangelical apocalyptic conspiracists engage in the field of politics and attack epistemological institutions in power to establish their idea of truth. "Dispensationalists are not the fatalists they propound to be" (Sturm, 2010: 133) and still believe they can "change history" (2010), for instance, through the endorsement of apocalyptic conspiracist discourses which can result in concrete political action, as prominently illustrated by the Storming of the American Capitol in January 2021 (Bond and Neville-Shepard, 2021). Several Christian symbols were displayed (e.g., Bibles or signs with Christian messages) during the Capitol riots (Schor, 2021). While the exact proportion of evangelicals or dispensationalists who took part in the Storming of the Capitol cannot be precisely determined, Opderbeck (2022: 587) observes an "enormous influence of dispensational end times theology" on the conspiracist belief of a rigged 2021 presidential election as "Donald Trump is viewed as a kind of last-chance bulwark against the apocalyptic forces of the Antichrist" (2022). Consequently, evangelical apocalyptic conspiracism should be discussed as a call to political action, and its discourses can influence American geopolitical practices through political networks of power. Furthermore, they have the ability to influence the political worldview and environmentally relevant behavior of American evangelical Christians.

As knowledge and power are intertwined, as we discussed in Chapter 2, the generation of evangelical apocalyptic knowledge contradicts the knowledge-creating strategies of the epistemic authorities. An important epistemological characteristic of evangelical apocalyptic conspiracism, and one that differentiates American evangelical knowledge-claims from their non-religious counterparts, is the legitimation of discourses through the use of biblical authority. By citing this fundamental source of knowledge for American evangelicals (Worthen, 2014; Bebbington, 1993), conspiracist knowledge-claims are seemingly legitimized for individuals with similar acceptance dispositions (Goldman, 2002; Fricker, 1998), which in this case is premillennial eschatology and the expectation of the rise of the Antichrist. Members of the social bonding group of premillennial evangelicals likely trust knowledge-claims which are communicated as divine truth, derived from the Bible, since trust in biblicism enacts what Shapin (1994: 7) calls a "moral bond" (Shapin, 1994: 7) between communicators and recipients of knowledge-entailing discourses. Still, certain scientific philosophies accompany the legitimization of apocalyptic conspiracist knowledge-claims, as evangelicals do not reject science but "have a long tradition of believing that religious and scientific methods will produce the same truths" (Evans, 2011: 712). Therefore, evangelical apocalyptic conspiracist discourses also employ the authority of science to further legitimate knowledge but usually only cite those scientific studies which support the desired truth-claim, as the confirmation bias or process of motivated reasoning affects what is perceived as valid scientific sources (Barkun, 2013; Aupers and Harambam, 2018).

Conclusion: The Greening of Theology

This chapter dealt with the political dimension of evangelicalism as well as the interdependencies between American evangelical religion and environmental attitudes. It is important to acknowledge the impact religious doctrine can have on the environmentally relevant beliefs and practices of believers. Gerten and Bergmann (2012) note that religions can be crucial driving forces of environmentally relevant behavior, and Hulme (2015a) further identifies religions as a cultural resource that has an impact on how humans perceive climate change or think about the environment more generally.

Still, religious doctrine is not the sole determining factor in adherents" perception of climate and environmental change, but instead it is the entire cultural and political belief system and overall societal values that have a bearing on climate change attitudes. Livingstone (2002b: 350 italics in original) notes that Eastern religions, which traditionally teach a close relationship between humans and their natural environment, but that "their *practices* were every bit as destructive as those in the West." Doctrine and beliefs are not the deciding factors in a society's relation to the natural environment, but the entire assemblage of religious, political, and societal dispositions that shape practices needs to be considered. Hence, we introduced four key characteristics that can be identified in apocalyptic conspiracist discourses and their political and cultural implications. Consequently, in the next chapter, we will not only discuss premillennial attitudes in terms of eschatological doctrine but also different political and societal dispositions that are embedded in evangelical premillennialism to explore how apocalyptic conspiracism is part of a collective interplay of different yet interconnected affects and perceptions of climate change.

5

The Construction of Evangelical Apocalyptic Conspiracist Climate Change Discourses

Since the rise of Christianity in Europe, apocalyptic discourses have argued for the imminence of the biblical End-Times and the approaching Return of Jesus Christ to Earth (Collins, 2020). The first verse of Revelation announces "things which must shortly come to pass" (Rev. 1:1, KJV). It is a core feature of Christian apocalyptic eschatology to proclaim the imminence of the final judgment "in which good and evil will finally receive their ultimate reward and punishment" (O'Leary, 1998: 6). American evangelical premillennialism is no exception. Through American evangelical history, several events and global developments were perceived as signs of an approaching apocalypse, including but not limited to the American Civil War, the World Wars, the establishment of the Israeli state, and the Cold War (Hummel, 2020; Sutton, 2017; Boyer, 1992).

In 1970, the popular evangelical prophecy writer Hal Lindsey concluded his best-selling non-fiction book of the 1970s, *Late Great Planet* Earth, with the term *Maranatha*, which, according to Lindsey, means "the Lord is coming soon." Lindsey (1980: 188) claims that "the time is short" as he sees several geopolitical events signaling Armageddon, the Rapture, and the Return of Christ. "For Lindsey, the 1967 Six-Day War and the seizure of Gaza, the West Bank, and Jerusalem, affirmed to him that God had set in motion an apocalyptic timeline" (Sturm and Albrecht, 2021b). Refraining from any explicit date-setting methods, Lindsey argued that the End of Time will occur within a forty-year-long Bible generation from the establishment of the Israeli state in 1948 (O'Leary, 1998). Lindsey strongly suggested that the eschatological belief expressed through *Maranatha* should have been fulfilled in or before 1988.

Fast forward to the year 2015, Hal Lindsey, now also known as an "intervangelist" (Bekkering, 2011), communicates his prophecies via his TV show *The Hal Lindsey Report*,[1] which is also distributed digitally through his personal website, hallindsey.com. Regardless of past prophecies which failed to unfold, Lindsey remains faithful to his End-Times message and claims that "we need to recognise that Christ [is] coming for us . . . [and] the Rapture is imminent" (Lindsey, 2015a). On the February 27, 2015 episode of *The Hal Lindsey Report*, Lindsey again sees evidence for the imminent End-Times, this time in the preparations for the United Nations Climate Change Conference (COP21), the aim of which was a legally binding international treaty to combat anthropogenic climate change. Lindsey (2015a) says that "Satan's lie of man-made climate change" will be used to centralize power and establish the Antichrist's global dictatorship that precedes Jesus' return to Earth and the subsequent millennium. Echoing established American premillennial dispositions, which expect that the Antichrist will rule over Earth during the Tribulation, Lindsey perceives climate change as a hoax, a satanic deception that the UN uses to establish the Antichrist's one-world-government under the pretext of environmental protection. In short, for Lindsey, the UN's efforts to reduce global greenhouse gas emissions through an international binding treaty confirm the imminence of an apocalypse.

In this chapter, we will explain why climate change and related political environmental discourses strengthen the belief in the imminent Return of Jesus to Earth and how evangelical apocalyptic conspiracism affects the perception of climate change. Our argument centers on a reciprocal relationship between climate change and preexisting socially accepted beliefs in the field of American evangelical apocalypticism. We will show that historically established eschatological, (geo)political, cultural, and conspiracist beliefs affect the perception of climate change and environmental politics, while conversely, climate change and environmental politics are framed in a manner that reinforces these preexisting apocalyptic conspiracist beliefs. In particular, the apocalyptic geopolitical imagination of the expected Antichrist's one-world government drives much of the suspicion toward global climate change policies. Climate change is perceived as a pretext for justifying the establishment of a suppressive world government as a solution under the pretext of climactic and environmental crises.

Epistemologically, we approach apocalyptic conspiracism as a self-sustaining system and explain how the "lens" of evangelical apocalyptic conspiracism affects the perception of climate change and related fields and how, vice versa, knowledge about climate change becomes portrayed in a manner that confirms the trueness of evangelical premillennial beliefs.

We will first outline key features of evangelical apocalyptic conspiracist discourses on climate change and elucidate how premillennial conspiracist climate change denial is distinct from other forms of American evangelical anthropogenic climate change skepticism (ACC-skepticism) while highlighting a relation between both discursive fields. In the second section, we engage with the reciprocal relationship between the fields of evangelical apocalyptic conspiracism and climate change, environmentalism, and climate politics. Here, we argue that preexisting apocalyptic worldviews influence the perception of climate change and related discourses. Conversely, knowledge on climate change and environmental politics becomes adjusted to the preexisting acceptance dispositions to reconfirm the imminency of the End-Times as well as preexisting evangelical apocalyptic imaginations of evil (*Feindbilder*). Evangelical conspiracist End-Times thought constitutes a resistance to the dominant power-knowledge system of anthropogenic climate change. In that context, evangelical apocalyptic conspiracist authors challenge epistemic institutions in power and contribute to increasing the influence and power of conspiracy theories and science skepticism. Therefore, in the third and final section of this chapter, we explain how authors portray themselves as the suppressed critics of the dominant knowledge-power system of anthropogenic climate change and how authors attempt to delegitimize the specious propaganda of climate change.

Section I: Blurred Boundaries between Apocalyptic Conspiracism and Skepticism

Hal Lindsey is likely the best-known voice linking climate change with End-Times beliefs, but there are many other authors, bloggers, self-proclaimed prophecy interpreters, YouTubers, and podcasters from the evangelical digital sphere who publish discursive material that interprets global warming, climate change, or environmental politics as signs of the rise of the Antichrist and the biblical End-Times. There are several variations of premillennial conspiracy thought, each with distinct underlying beliefs and motives that affect the construction of the particular text. In that context, it is important to differentiate between what we call evangelical apocalyptic conspiracism and non-conspiracist premillennial discourses on climate change, or evangelical ACC-skepticism.

In line with the definition of evangelical apocalyptic conspiracism described in the previous chapter, discourses of interest for this study must claim to unveil knowledge about the rise of an apocalyptic evil that uses deception or a superconspiracy to establish a New World Order that will eventually usher in the biblical End-Times, including the Rapture, Tribulation, and Jesus' Second Coming. In this section, we illustrate that not all evangelical discourses which reject the science of anthropogenic climate change, climate protection policies, or environmentalism fall into the category of apocalyptic conspiracism. Nevertheless, it is important to stress that there are blurred boundaries. Since no piece of evangelical discourse that questions the dominant power-knowledge system of climate change can emerge isolated from preexisting evangelical discourses, intertextual relationships between premillennialism, conspiracism, and ACC-skepticism must be considered.

Non-Apocalyptic Conspiracist Evangelical ACC-Skepticism (and Its Conspiracist Tendencies)

Not all American evangelical premillennial or apocalyptic discourses which address climate change, global warming, or environmental degradation necessarily relate to a conspiracy that will precipitate the reign of the Antichrist. Rather, the physical indicators of climate change, global warming, environmental degradation, and natural disasters are frequently interpreted as the result of divine will and indicators of the biblical End-Times. Zaleha and Szasz (2015) argue that the general, not necessarily conspiracist, premillennial expectation of the imminent End of Times and Jesus' Second Coming lowers the desire to protect the environment because climate and environmental change are seen as a proof for the apocalyptic decay of Earth. According to a study by Yale University and George Mason University, about 26 percent of self-identified evangelical and born-again Christians in the United States see no need to worry about global warming because the End-Times are coming (Roser-Renouf et al., 2016). Even though such evangelical End-Times beliefs do not automatically deny the existence of anthropogenic climate change, expectations of the imminent end of the world can fuel the resistance toward global warming policies (Barker and Bearce, 2012). Carr et al. (2012: 292) write that some participants in qualitative interviews at churches in Texas "felt there was no need to worry because this world would end before climate change had any serious impacts," even though this was "distinctly a minority view" compared to apocalyptic conspiracist attitudes. An example of a non-conspiracist premillennial response to climate change is presented in the book *Global Warming or God's Warming* by the evangelical prophecy writer Desmond Michael Coverley. Coverley (2017: 351), one of the only Black authors we are aware of within this apocalyptic evangelical milieu, argues that "global warming and climate change is an instrument that God is using to warn people of the last day's preparation." Coverley further speculates that it could always have been God's plan that humans destroy the planet through greenhouse gas emissions and, moreover, that climate scientists are advised by God to teach about the future of the planet.

In short, there are several non-conspiracist American premillennial responses to climate change that interpret climate change or global warming as a real phenomenon. Some assume that climate and environmental change are caused by God to herald the apocalypse, whereas other premillennial responses are potentially compatible with mainstream climate science of anthropogenic climate change. Still, those "apocalyptic narratives in US Evangelicalism have been implicated in political inaction on climate change" (Jenkins, Berry, and Kreider, 2018: 96) because of this deterministic construct of an imminent apocalypse which cannot be avoided. Nonetheless, due to the absence of conspiracist knowledge-claims or embedding of climate change into apocalyptic Manichean discourses, such discourses do not fall into the category of evangelical apocalyptic conspiracism.

Moreover, it needs to be noted that "by no means do all sceptics advance conspiracy theories" (Douglas and Sutton, 2015: 99), a relevant observation which also applies to the field of American evangelicalism more generally. The rejection of knowledge-claims which argue for the existence of anthropogenic climate change and the necessity of reducing greenhouse gas emissions only falls into the category of conspiracy theorizing or conspiracism if it is argued that these knowledge-claims are *intentionally* faked to deceive the public deliberately for the benefit of some individuals or organizations (Soentgen and Bilandzic, 2014). Here, a distinction needs to be drawn between climate change conspiracy theories which address individual plots, like a "conspiracy among scientists to maintain or increase funding opportunities" (Sharman, 2014: 159) and the "full package deal" (Robertson, Asprem, and Dyrendal, 2018: 4) of conspiracism in which climate change is believed to be a "myth spread by environmentalists in the pursuit of a one-world socialist government" (Douglas and Sutton, 2015: 99).

The non-conspiracist evangelical rejection of anthropogenic climate change knowledge can be motivated by a belief in the sovereignty of God who would not allow humans to damage his creation, political conservatism, or by a rejection of "post-normal" climate science (Ecklund et al., 2017; Lowry, 2016; Evans and Feng, 2013; Carr et al., 2012). However, in some cases, there is no simple distinction between ACC-skepticism, conspiracy theorizing, and conspiracism. A good example of these blurred boundaries is the Cornwall Alliance, an evangelical ACC-skeptic organization that is frequently discussed in the academic literature (e.g., Gough, 2021; Ronan, 2017; Kearns, 2014, 2012; Zaleha and Szasz, 2015, 2014; Roberts, 2012; McCammack, 2007). This body of literature argues that the Cornwall Alliance typically employs a mix of biblical and scientific authority to claim that the climate does not significantly change due to anthropogenic greenhouse gases. In that context, the Cornwall

Alliance states that the Bible, in particular Gen. 1:28,[2] should be understood as a clear command to subdue and rule the Earth, as the non-human creation is supposedly designed to serve human needs, while it is further argued that the IPCC climate science is based on false assumptions. Often, evangelical ACC-skeptics criticize the predictive post-normal science of the IPCC but instead argue for a Baconian approach to science that contradicts many aspects of mainstream climate science. Conspiracy theories concerning an alleged intentional distribution of false knowledge-claims by climate scientists are typically not addressed in the above-mentioned academic literature on the Cornwall Alliance.

Some more recent articles posted on the Cornwall Alliance's website hint at underlying conspiracist beliefs, which possibly fuel some of the organization's ACC-skepticism. In an April 2020 blog post, the Cornwall Alliance's spokesperson Calvin E. Beisner (2020a) writes that regardless of possible conspiracy beliefs, the organization attempts to gain epistemic authority from evidence-based arguments rather than conspiracy accusations:

> I don't know how science could lie or conspire. Science isn't a personal agent. Scientists are. Do some lie? No doubt. Are some involved in some sorts of conspiracies? Probably. But far from accusing scientists of lying and conspiracy, we've self-consciously refused to do that, because we think the proper ground of argument is evidence and logic, not *ad hominem*.

Irrespective of this honorable promise to lead a seemingly objective discussion on climate change, the Cornwall Alliance's website contains some conspiracist knowledge-claims which lean toward the delegitimization of climate change protection policies and the environmental movement as a whole. Rainer Zitelmann (2021) states in a guest column on cornwallalliance.org that "anti-capitalists like [Naomi] Klein are only superficially concerned about the environment and climate change. Their real goal is to eliminate capitalism and establish a state-run, planned economy." In another article on the website by H. Sterling Burnett (2021) titled "Climate Alarmists Call for Global 'Eco-Dictatorship'," the author claims that "the underlying goal of leading climate alarmists, if not their unthinking minions protesting in the streets and at schools, is really about imposing socialism on masses who have repeatedly rejected and overthrown it in the past." Several other Cornwall Alliance articles at least hint at the possibility that climate change is connected to a planned profound transformation of the world's political and economic systems, which makes them not only conspiracy theories but conspiracists: the holistic belief in a global superconspiracy that not only assumes some individual conspiracies but that human destiny on a global scale is controlled by evil cabals (Robertson, Asprem, and Dyrendal, 2018;

Barkun, 2013). Furthermore, accusations concerning an intended "eco-dictatorship" fall, to some extent, into the category of apocalypticism, as it is argued that the true purpose of climate change is *revealed*: the intended abolishment of established power structures in favor of a rather sinister global dictatorship.

Following the definitions we have established, important characteristics of evangelical apocalyptic conspiracism are absent from the Cornwall Alliance's pronouncements. Most importantly, the Cornwall Alliance does not clearly advance premillennial eschatology. Although Zaleha and Szasz (2014) argue based on Beisner's statement in a 2006 interview[3] that the Cornwall Alliance's spokesperson's ACC-skepticism is driven by his premillennial eschatology, we could not identify any clear indicators of premillennial discourses on the Cornwall Alliance's website or other publications. There are no blog posts on cornwallalliance.org we are aware of that mention the Antichrist, the Rapture, Jesus' Second Coming, imminent End-Times, or other indicators of American evangelical premillennialism. Additionally, Beisner's recommendation of books on amillennialism indicates that he does not follow a premillennial eschatology.[4] Furthermore, the Cornwall Alliance does not argue in a deterministic manner that the End-Times or establishment of a global (eco-)dictatorship is imminent and cannot be stopped. For example, in the above-mentioned article on the Global Eco-dictatorship, Burnett (2021) hopes that "free people make the right choices in the coming elections. If not, we will lose our freedom, our prosperity, and environmental quality." This evident acknowledgment of human agency contradicts the deterministic, fatalistic, and prophetic features of most evangelical premillennial discourses.

The prominent example of the Cornwall Alliance illustrates the complexity of American evangelical ACC-skepticism, as well as the blurred boundaries between different forms of skepticism, apocalypticism, and conspiracism. Concerning these blurred boundaries, it is important to keep in mind that knowledge travels between different digital spaces and fields of evangelical ACC-skepticism. For example, in a 73-minute-long YouTube video on the "Prophetic Significance of US Withdrawal from Paris Climate agreement," the Texan pastors Andy Woods and Jim McGowan (2017) refer to publications by Calvin E. Beisner to support their conspiracist discourses, which see Trump's withdrawal from the Paris Agreement as a "Setback for Satan." In the video, the pastors utilize Beisner's evangelical ACC-skeptic knowledge-claims, which state that climate science is false and that anthropogenic climate change cannot be real due to God's sovereignty, to support their worldview. Conversely, as mentioned earlier, the Cornwall Alliance slowly began to employ conspiracist knowledge-claims in recent years. These interdependencies illustrate that the superordinated objective of delegitimizing anthropogenic climate change knowledge can be more important than distinct religious, eschatological, and cultural beliefs that are typically communicated by the Cornwall Alliance and apocalyptic conspiracists like Woods and McGowan (2017).

Section II: The Reciprocal Relationship between Climate Change and Evangelical Apocalyptic Belief

By analyzing texts that emerge from digital evangelical apocalyptic spaces, we illustrate in this section that apocalyptic conspiracist authors frame climate change and related politics in a manner that reconfirms established apocalyptic truth-claims while, conversely, the perception of climate change is affected by preexisting evangelical premillennial conspiracist dispositions. Developments of global importance, such as climate change, are not only used to reconfirm the premillennial belief in the imminency of the Rapture, Tribulation, and Jesus' Second Coming. Rather, climate change is employed to validate the entire evangelical apocalyptic conspiracist worldview, which not only consists of premillennial doctrine but also interconnected religious, political, social, and cultural beliefs as well as established *Feindbilder*. We present two domains that illustrate the reciprocal relationship between climate change and preexisting evangelical apocalyptic truths. First, we engage with the political and economic framing of climate change and show how the political responses to climate change are used to reconfirm the antiliberal bias of evangelical apocalypticism. Second, we illustrate how the environmental movement is perceived as the rejection of Christ in the Church Age, the final disposition before the apocalypse. As evangelical apocalypticists expect that humanity will reject Christ and "true" Christian teachings before the apocalypse, authors construct environmentalism as being a false religion to confirm this apocalyptic expectation. In that context, the papal environmentalism of Pope Francis is used to reinforce suspicions toward the Catholic Church and the rise of a false religion and false prophets. Certainly, all of the following elaborations must be seen in the context of a particular anti-globalist apocalyptic geopolitical worldview, which drives much of the evangelical apocalyptic conspiracist resistance to climate change.

The Role of the State and Evangelical Apocalyptic Anti-liberalism

Political and economic discourses are dominant and recurring themes in debates about climate change, and the field of evangelical apocalyptic conspiracism is no exception. Climate change frequently acts as a proxy for political and socioeconomic conflicts about the role of the state and acceptable levels of governmental regulation (Turnpenny, 2012). Climate change is not just science, isolated from the rest of society, but a form of knowledge that implies political action (or inaction). Conservative American evangelicals are more likely to oppose established scientific knowledge if it implicates the necessity for political and social change or when it initiates a moral competition between science and religion (Evans and Feng, 2013; Kearns, 2012; Evans, 2011; Noll, 2002). In that context, Kearns (2014: 167) describes "the economic framing" as the "real battleground" of American evangelical climate change debates. Overall, a "historical intertwining of evangelical identity, conservative politics, and climate change skepticism" drives much of the American evangelical resistance to climate and environmental protection policies, a connection that can also be observed in American evangelical apocalyptic conspiracist digital spaces (Sheldon and Oreskes, 2017: 363). The majority of articles contain some references to the socioeconomic and (geo)political implications of climate change, typically arguing against any form of taxation, restrictive climate change politics, liberalism, or socialism while endorsing capitalism and libertarianism. Such arguments are embedded in an apocalyptic conspiracist framework in which restrictive crisis politics, governmental regulation, or the redistribution of wealth are not simply considered as measures to mitigate the adverse effects of climate change but as the actual purpose of climate change.

Conservative American evangelicals view "free enterprise capitalism as God's way and the natural order of things" (Amos, Spears, and Pentina, 2016: 7). Within evangelical apocalypticism, liberal, socialist, or communist policies are usually connected to the Antichrist's global government, so that any issues and policies that could indicate increased regulation or taxation are framed as being an expression of apocalyptic evil. It is often assumed that globalist or foreign forces threaten to impose socialism/communism onto the United States through the backdoor of climate change as a pretext for increased (global) governance. For many apocalyptic conspiracists, taxation does not only mean loss of money but loss of power, freedom, and the rise of the evil. On the blog EndTimesBibleProphecy (no date) it is written that solutions to climate change "typically involve massive tax hikes, loss of national sovereignty, and a massive transfer of wealth and power to new global institutions. This "crisis" is how the globalists hope to get nations (and the United States in particular) to cede power to them." Also addressing global climate protection policies, RaptureReady's Geri Ungurean (2016) denotes the UN's Sustainable Development Goals as "Marxism in its purest form." While these utterances on socialist/communist endeavors in relation to climate change must be discussed in light of anti-globalist attitudes and the vestiges of the Cold War (see Chapter 7), not all authors directly address globalism in relation to the economic and political framing of climate change. Addressing the perceived loss of control through increased governance in an interview with the Lamb & Lion Ministries/Christ in Prophecy, Andy Woods claims that "if they can convince us that the climate is in a state of crisis then we will accept pervasive government controls over virtually every area of life" (Lamb & Lion Ministries, 2016b). In another example, EndTime Ministries' Dave Robbins claims that climate change is the newest attempt to destroy the capitalist system of the United States. Robbins (2019a) claims that "the Dirty word is now carbon rather than capitalism" and that "wealth redistribution is exactly what carbon trade and tax laws are designed to accomplish."

Apocalyptic conspiracist authors and websites refer to the "Green New Deal" (GND) as an idea that intends to restrict the freedom of Christian Americans through environmental regulation. As a concept of a series of governmental programs and interventions that aim to initiate a transition toward a fairer and more sustainable society, the GND gained prominence in the United States as a resolution introduced in early 2019 by US congresswoman Alexandria Ocasio-Cortez and Senator Ed Markey. Galvin and Healy (2020: 1) comment that the GND's "chief aims are to radically decarbonize the US economy while significantly reducing economic inequality, in such a way that these two achievements would be inextricably linked, and the rights of vulnerable communities protected and enhanced." The GND's additional aim of reaching net-zero greenhouse gas emissions by 2030 was criticized as too ambitious by many US politicians, even by Democrats who were generally supportive of the GND. While many Republican politicians referred to the GND as a form of socialism, Joe Biden eventually adopted many proposals of the GND in his administration's climate plan, which was announced in April 2021. Yet, the GND is not just an American concept; many states and intergovernmental institutions work toward a more sustainable and fairer economy while employing the terminology of a Green New Deal as homage to Roosevelt's 1930s New Deal.

In an article called "The Devil is Using the Democratic Party to Destroy America" on RaptureForums, Larry Tomczak (2019) writes that the GND is an attempt by the Democratic Party to introduce socialism in the United States as the climate crisis is "hysteria not based on scientific empirical evidence but much from Al Gore's theories and disproven myths catapulting him to multimillionaire status." On the Lamb & Lion Ministries' online show Christ in Prophecy, Brandon Holthaus (Lamb & Lion Ministries, 2022) states that the GND is just another repackaging of other conspiracist plots like the UN's Agenda 21 or Agenda 30, as the GND must be perceived in a holistic manner as part of the Great Reset that prepares the world for the Antichrist. Similarly, the website Z3 News puts the GND into the discursive field of the Great

Reset and the Covid pandemic. According to the journalist Matthew Ehret (2021), whose original article was also published on Z3 News, "the IMF, World Bank, UK, USA, corporate and banking sector to take advantage of COVID-19 to shut down and 'reset' the world economy under a new operating system entitled the Green New Deal." Ehret's article not only illustrates the holistic and plastic characteristics of apocalyptic conspiracism that can easily absorb and incorporate leading international economic and political institutions as well as events like Covid, but also how evangelical apocalyptic conspiracism resonates with the secular conspiracist right.

The original article by Ehret was published outside the realm of evangelical apocalypticism, the networked posting and reposting in the digital environment added new meaning to the same piece of discourse. As all kinds of discourses "are located in a world filled with prior utterances and are therefore implicated in an implicit dialogue with that pre-populated world of discourse" (Hodges, 2015: 43), the non-evangelical apocalyptic conspiracist text automatically refers to previous discourses in the digital space that assume an anti-Christian superconspiracy. Consequently, the meaning of a text changes depending on the digital spaces in which it is consumed. Due to an intertextual relationship with previous evangelical apocalyptic discourses, Ehret's (2021) text reconfirms the evangelical apocalyptic conspiracist worldview on the website Z3News.com without mentioning any concrete biblical End-Times belief. Such an example highlights what Livingstone (2005b: 399) calls the "cultural geographies of reading." Livingstone (2005b) argues that the social and cultural composition of a certain space plays a crucial role not only in the production of knowledge but also in the reception of knowledge. Depending on established beliefs, knowledge, and truth-claims of the space in which a consumer of knowledge reads a text, different meanings are attached to the text. As "knowledge usually does not move around the world as an immaterial entity," the meaning of texts changes while circulating through space (Livingstone, 2005b: 391). And in the example discussed here, evangelical apocalyptic meaning gets attached to a non-religious, anti-globalist right-wing discourse. Generally, it is common practice in American evangelical apocalyptic conspiracism to embed non-religious or non-apocalyptic texts into evangelical digital spaces to assign a new meaning to the text that confirms preexisting religious apocalyptic knowledge.

However, the evangelical apocalyptic conspiracist objection to the GND echoes resistance to the original New Deal, while the established *Feindbild* of the regulating political left is reinforced. The New Deal was a series of liberal regulations, financial and social reforms, and public work projects introduced by the former president Franklin D. Roosevelt in the 1930s. As a response to the Great Depression and the increasing unemployment rates in the early 1930s, the Roosevelt administration introduced an unprecedented series of governmental interventions that many Americans perceived as a restriction of personal and economic freedom.

Addressing the evangelical apocalyptic response to the New Deal, Sutton (2012: 1061) found that "white fundamentalists across the continent came to believe that New Deal liberalism was the means by which the United States would join the legions of the antichrist." Since "premillennialism served as the filter through which the faithful understood American politics" (2012), apocalyptic evangelicals of the 1930s (described as fundamentalists during this period by Sutton, 2017, 2012) began to connect premillennial prophetic beliefs of evil totalitarian powers to real-life politicians and their policies. In constant anticipation of the End-Times, many apocalyptic evangelicals perceived Roosevelt's increased governmental regulation as an overreach of power and thus a sign of the rise of apocalyptic evil. Certainly, the emerging relationship between evangelical End-Times beliefs and Roosevelt's New Deal must be discussed in relation to the general political attitudes of conservative evangelicals.

Sutton (2017, 2012) shows that many apocalyptic evangelicals themselves are politically active citizens of the United States and therefore preach an assemblage of beliefs and truths in which political arguments are eschatological, or in which eschatological arguments are political—an observation that is still relevant. Sutton (2012) further argues that some evangelical End-Times preachers even discussed Roosevelt as a potential candidate for the Antichrist, which is a remarkable geographical shift in American apocalyptic imagination. Before the 1930s, it was commonly assumed that the Antichrist will rise to power within the boundaries of, or close to, the Roman Empire (see Chapter 7). Leaders like Stalin, Hitler, or Mussolini were also discussed as possible incarnations of the Antichrist (Sutton, 2017).

As the apocalyptic conspiracist resistance to the GND shows, the interpretation of the New Deal as a sign of the End-Times had a long-lasting impact on the field of American evangelical apocalypticism. Sutton (2017: 254) writes concerning evangelical apocalyptic responses to the New Deal that "by preaching cultural engagement and working for conservative political causes while simultaneously predicting an imminent Armageddon, fundamentalists made subsequent generations of believers aware of the dangers of modern liberalism." Yet, the anti-liberalism of the 1930s drew from historically formed attitudes of American evangelicalism and Protestantism more broadly, famously observed in Max Weber's *The Protestant Ethic and the Spirit of Capitalism* (1905). For instance, Marsden (1980) observes that in the nineteenth century, a link between free enterprise capitalism and evangelicalism was well established, as many evangelical Protestants believed that the pursuit of wealth was a legitimate and even providential aspiration in harmony with morality, and as such, increased governance could hinder cultural progress and the perception of the United States as a Christian nation.

Still, according to Sutton (2017, 2012), it was in particular the evangelical apocalyptic resistance to the New Deal in the 1930s that established eschatologically driven anti-liberalism in the American evangelical culture. Werly (1977) and Herman (2001) further observe that in the twentieth century, the evangelical apocalyptic objection to liberalism and governmental regulation was constantly reinforced by applying established imaginations of evil to contemporary politicians and political institutions that promote liberal politics. Contemporary evangelical apocalyptic conspiracist resistance to climate protection policies and the GND needs to be understood as a practice that reconfirms the truth of the nefarious thoughts and actions of communists, socialists, or liberals who attempt to restrict the freedom of Americans through taxation and regulation. "The protestant commitment to individualism and freedom" observed by Hofstadter (1964: 80) is ever-present among evangelical apocalyptic discourses and illustrates highly sustainable political and economic attitudes among evangelical apocalyptic discourses. New developments like climate change politics and the GND stabilize apocalyptic conspiracist belief systems as long as core religious, cultural, and political assumptions are not questioned (Howard, 2006). In that context, political and economic debates concerning climate change are framed as being pivotal for the apocalyptic conflict between good and evil, in which any form of support for governmental regulation, liberal politics, and restrictive climate protection policies is portrayed as paving the way for the Antichrist. Thereby, the resistance to climate protection policies or carbon taxes is portrayed as a moral imperative in which any interdependencies between economic practices and carbon emissions become irrelevant as the only objective is to prevent the redistribution of wealth and thus the introduction of anti-Christian global socialism/communism under the guise of a sustainable economic transformation.

The Church Age, the False Religion of Environmentalism, and Papal Environmentalism

Prophetic truth is also reconfirmed through climate change via a perceived rejection of Christ and rise of false prophets and religion in the final dispensation before the apocalypse. According to dispensational premillennial eschatology, humanity currently lives in the Church Age, a dispensation in which humanity will reject Christ to follow a false religion and false prophets (Gribben, 2011; Boyer, 1992). Most prominently popularized by the "social pessimism" (Gribben, 2011: 85l, 2024) of John Nelson Darby, apocalyptic evangelicals expect the downfall of humanity in the current dispensation that is connected to the rise of a false religion.

The association of the climate movement with the Antichrist was observed as early as 1992. Hopewell (1998) describes a 1993 sermon where John Hagee—the founder of Christians United for Israel and whose endorsement for presidential candidate John McCain was rejected because of anti-Catholic views—takes issue with a newspaper clipping of the Earth Summit in Rio de Janeiro depicting "egregious" environmentalists sitting in a circle, a scene that did not escape Hagee's keen eye for the signs of the times. Hagee preaches, "I recognized the significance of the magic circle used in satanic and occult worship" (Hopewell, 1998: 66). However, Hagee is not completely unsympathetic to the environmentalist's mission, he too wants clean air and water, but he goes on to say that we should not be fooled: the environmentalist movement simply employs these tactics to vitiate and establish global dominance and evil: "[T]he environmental movement is not about conservation. It's about creating an environmental juggernaut that marries the New World Order crowd and the New Age occultists with the objective of bringing about a global crisis that can be solved by a one-world government". Hagee continues his admonition, "[t]hose who went to Rio opened themselves, willingly or unwillingly, to the invasion of demon spirits". Satan's secret plan has then been debunked in the Bible, and the only question concerning the environment has already been answered in the Bible as part of the End-Time events: "The earth's atmosphere will be destroyed in a nuclear explosion. That will be the end of the environmental issue. And there's not one thing you can do about it." The "pagan cretins," as Hagee terms environmentalists, will finally be "exterminated".

An important feature of apocalypticism is being granted a special or elect knowledge of truth about the past, present, and future, and it assumes that the majority of humans will be deceived by false prophets (DiTommaso, 2020). The environmental movement is often imagined as such a false religion, a kind of earth worship, and it confirms that humanity currently lives in the Church Age (Herman, 2001; Spark, 2001). Following this, evangelical apocalyptic conspiracist authors engage with climate change not only as a form of politically relevant scientific knowledge but also as a social movement that surrounds debates about climate and environmental change. In short, it is argued in such digital spaces that environmentalists and advocates of climate change protection policies propagate the worship of Earth instead of the worship of God, and that thereby, the prophesied rejection of Christ during the Church Age is fulfilled.

Addressing the perceived decline of "true" Christianity, RaptureReady's Mark Becker (2021) writes that "as Christians, we shouldn't be startled by these things, as we live in a fallen world," as he ascertains that "the One World Religion is also using 'climate change' to unite the world's religions that will be dominant until the midway point of the Tribulation." The website ProphecyNewsWatch (2019) observes the imminence of the End-Times where the "religion of climate change... has its unique gods (like Mother Earth)" and "high priests and religious leaders (the climate change gurus and radical environmentalists) as well as its patron saints (like Sweden's Greta Thunberg)." In another example, addressing the Earth Day celebration in April 2022, RaptureReady (2022) declares:

> The radical far-Left is not against worship, religion, or bowing down to deity, just as long as it's not the God of the Bible. In fact, Earth Day is when their darkness seems to shine the brightest, as they become the Romans 1 "fools" that worship the creature more than the Creator. Today the pagan crazies were out in full swing, paying homage to their dirt mother. Pope Francis is a Gaia worshipper, too.

Todd Strandberg (2005), who is the editor of RaptureReady.com and the "Rapture Index," a calculation of forty-five variables assigned points as to their effect on bringing about the Rapture on any given day, includes eleven environmental variables: Wild Weather, Famine, Drought, Climate, Global Turmoil, Earthquakes, Volcanoes, Food Supply, Floods, Plagues, and Oil Supply and Price. Although Strandberg is now a climate skeptic (see Chapter 7), in 2005 he explained his timeline in relation to global climate change: "I used to think there was no real need for Christians to monitor the changes related to greenhouse gases. If it was going to take a couple hundred years for things to get serious, I assumed the nearness of the End-Times would overshadow this problem. With the speed of climate change now seen as moving much faster, global warming could very well be a major factor in the plagues of the tribulation" (2005). The Rapture Index is not unlike the Doomsday Clock that was developed in 1947 by a group of scientists who believed the end would come about by nuclear holocaust . It is precisely these types of secular prophecies that normalize and stoke the premillennial fire with evidence that the end is near.

On RaptureForums, Terry James (2021) describes the biblical origins of warnings of the environmentalist "anti-God worship system." According to James (2021), the apostle Paul warned in Rom. 1:20-25[5] that people will serve "the creature more than the Creator" as "God is neither needed nor wanted in this totally evil religion [environmentalism]." Another important truth of evangelical apocalyptic conspiracism that is reinforced by climate change discourses is also addressed by James on RaptureForums (2021): "Catholic Pope is part of the wickedness."

Generally, the idea of a false religion is often applied to the Catholic Church and the pope, who is believed to be the person who establishes, or contributes to establishing, a false one-world religion through lies and deceptions. In particular, after the publication of the papal encyclical letter *Laudato Si'*, apocalyptic conspiracist writers perceived papal environmentalism as proof of his complicity (Graham, 2020; Iheka, 2017). In the encyclical letter, Pope Francis expresses his concerns about humanity's interaction with their natural environment while criticizing growth-based economies and consumerism. With taking on the name of St. Francis of Assisi, the Patron Saint of Ecology, Pope Francis sent "a strong signal that he would focus on the poor and had affectionate feelings toward non-human organisms" (Taylor, Van Wieren, and Zaleha, 2016: 307). Yet, this signal is also interpreted as signaling the End-Times.

In his persistent search for indicators of Jesus' Second Coming, Hal Lindsey (2015b) describes climate change on his website as an "intentional deception that our nation's leaders and many of the world's prominent leaders, including Pope Francis, are propagating." Through relating the Vatican and the pope to the Antichrist's deception (Boyer, 1992), Lindsey echoes established beliefs of the digital evangelical apocalyptic sphere, which hold that Pope Francis deliberately contributes to the establishment of the apocalyptic world government through publicly supporting environmentalism and international climate protection politics. Addressing Francis' environmental engagement in 2015, EndTimeMinistries' Dave Robbins (2015) argues that the pope could turn into a false prophet due to his environmental activism and support of the UN. Robbins (2015) writes that "along with climate change, Pope Francis promoted other United Nations global-governing platforms such as sustainable development and social justice." In the September 2015 issue of the online periodical the Lamplighter, the pope's 2015 visit to the United States is used to discuss the possibility of Francis being the Antichrist (Lamb & Lion Ministries, 2016a). Here, the authors of the periodical state that due to Francis' globalism and his supposed emphasis on global warming rather than Jesus Christ, he could be a possible Antichrist.

Pope Francis' environmentalist positionality is frequently discussed in light of his globalist endeavors and the allegedly desired establishment of a one-world religion. On Z3 News, Gal (2020a) writes that Francis' "support for global warming, open borders, equality and peace has earned Francis repeated praise and laud from the U.N." Generally, it is important to keep in mind the holistic characteristics of apocalyptic conspiracism, specifically the assemblage of different developments—the pope as promoter of globalism and socialism, ties to the UN as the Antichrist's institution, restriction of freedom through climate change policies, rise of a false religion of environmentalism—that confirm climate change as an apocalyptic superconspiracy. Therefore, evangelical apocalyptic conspiracist denunciations of Pope Francis' involvement in the climate change debate must be seen as part of a complex discursive assemblage that does not only attack the pope but also the political attitudes which evangelical apocalypticists associate with evil (globalism, liberalism, socialism, or communism).

Similar to the historically established antiliberal bias of American evangelical apocalypticism, climate change is used to reinforce preexisting dispositions and *Feindbilder*. For many apocalyptic evangelicals, it was a historically established presumption that the head of the Catholic Church is somehow related to the rise of the apocalyptic evil or that the pope himself would be the Antichrist. Yet, suspicion toward the pope is not merely an American evangelical phenomenon. Boyer (1992: 273) writes concerning the supposed relationship between the pope and the Antichrist that "no Antichrist tradition has a more ancient lineage than that pointing to the Pope" since, for example, "as early as the fourth century, Hilary, bishop of Poitiers, taught that Antichrist would arise from within the Church" (1992: 273–4).

In the American millennialism of the nineteenth and early twentieth centuries, the link between the pope and the Antichrist was rejected by many American prophecy writers, but after the Second World War, some premillennial prophecy writers and conspiracy theorists of the American right began to portray Catholicism as part of the apocalyptic evil and a threat to American Protestant freedom (Boyer, 1992; Werly, 1977). Several American evangelical apocalypticists in the twentieth century discussed links between the pope, Catholicism, and the rise of the apocalyptic evil, which led to a general dismissal of Catholicism in American evangelicalism (Sutton, 2017). Hal Lindsey "strongly hinted at a Pope = Antichrist connection" due to the increasing involvement of the Vatican in world politics (Boyer, 1992: 274). For Lindsey and many of his adherents and successors, like John Hagee, the geographical location of the Vatican indicates a link between the pope and the Antichrist, as the satanic force on Earth is frequently believed to lead a revived Roman Empire (Herman, 2000). This geographical imagination is also addressed by Robbins (2015), who writes:

> Remember that the ten horns on the world government beast symbolize the reborn Holy Roman Empire, currently known as the European Union. The Holy Roman Empire has always had two leaders—the most powerful political leader from Europe and the most powerful religious leader from Rome—the Pope.

Evangelical apocalyptic conspiracist discussions on the relationship between Pope Francis' climate advocacy and the rise of the Antichrist validate preexisting evangelical apocalyptic imaginations and the *Feindbild* in the pope or Catholicism.

Apocalyptic Conspiracism: A Self-Sustaining System

In one of the first engagements with the relationship between evangelical premillennial prophecy and conspiracism, Werly (1977) writes of influential American Christian nationalists of the early twentieth century (William J. Simmons, Billy James Hargis, Carl McIntyre, and William Dudley Pelley) and argues that these individuals contributed to popularizing religiously driven right-wing and *Feindbild* apocalyptic discourses. These Christian nationalists claim that free American evangelical Christians must resist the Antichrist-led Catholics and/or communists who seek to control the United States and the world through conspiratorial plots. Dualistic apocalyptic thinking and the combination of nationalism and faith are summed by Hargis' (ibid: 50) Manichean statement on America's relation to communism, where only two positions were possible: "Choose Christ, for Christ loves America" or "Communism, which is Satan."

These historically established rejections of socialism, communism, globalism, and Catholicism are ever-present within contemporary evangelical apocalyptic conspiracy discourses on climate change. Alternative climate change knowledge within evangelical apocalyptic digital spaces is shaped by preexisting premillennial and conspiracist beliefs that "function to reinforce the already held values of the community" (Howard, 2006: 42). In that context, the relationship between evangelical End-Times conspiracism and climate change needs to be discussed as reciprocal. Preexisting premillennial dispositions and imaginations of the truth affect the perception of climate change and lead to a rejection of "mainstream" climate change knowledge, while, conversely, the emerging alternative apocalyptic conspiracist climate change knowledge reinforces the previously held beliefs.

Therefore, evangelical apocalyptic conspiracism needs to be discussed as a non-falsifiable self-sustaining system (see Figure 1). Any global issue or crisis is often reframed in order to sustain the evangelical apocalyptic conspiracist belief system, whereas concurring information is portrayed as part of the conspiracy or, in the case of evangelical apocalypticism, as an anti-Christian deception. Moreover, this self-sustaining system illustrates that historically established truths that were formed and disseminated over the last centuries still matter in the digital world. Evangelical apocalyptic histories and knowledge formed in the physical world continue to affect the construction of knowledge and truth-claims in contemporary digital spaces because the "online" and "offline" do not exist separately from each other, as social relations transcend any binary conceptions. Adherents and authors of premillennial conspiracist belief systems create social connections through the use of language in an "online performance" of Christian apocalypticism (Howard, 2010: 731).

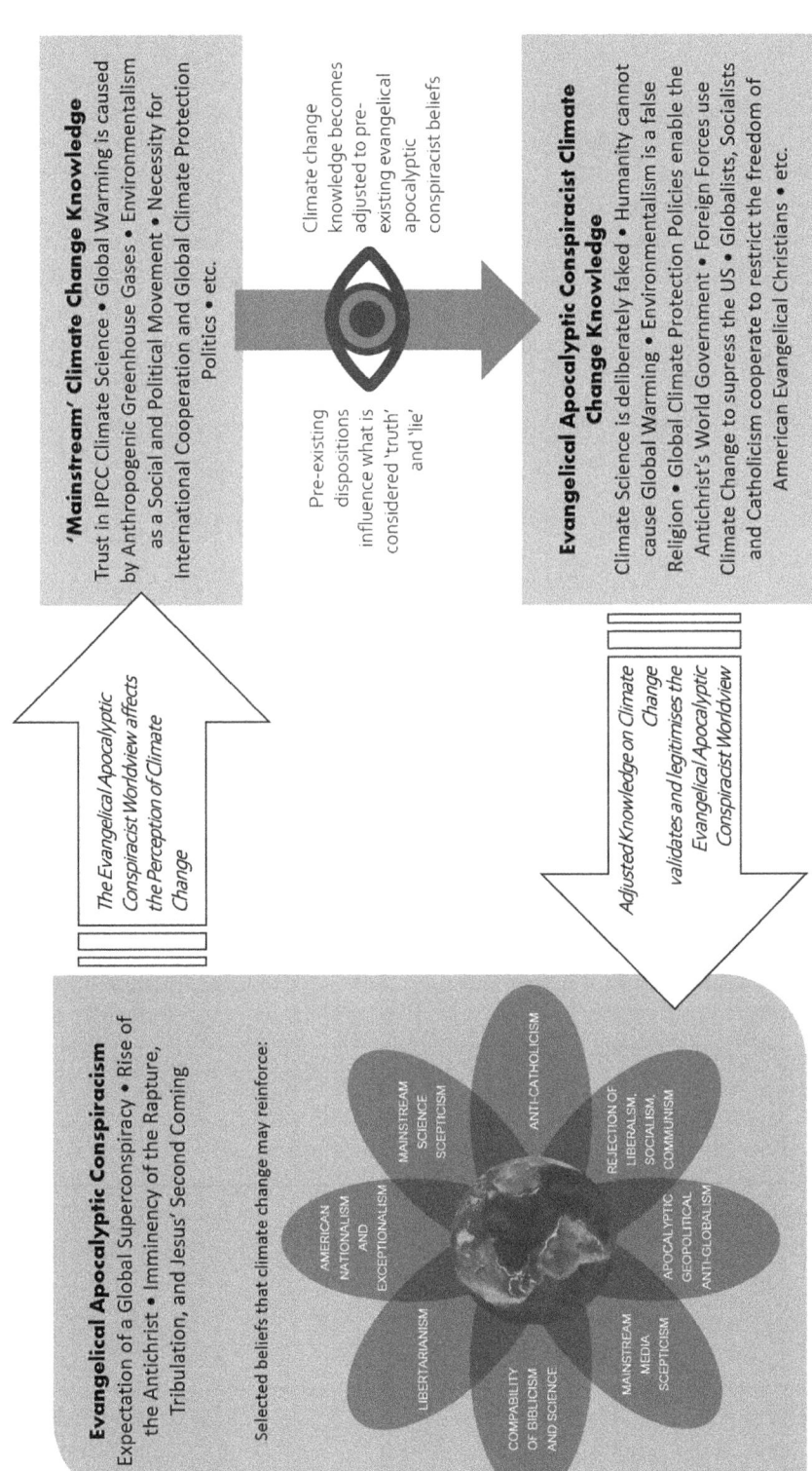

FIGURE 1 Reciprocal Relationship between Climate Change and Apocalyptic Conspiracism. Source: Authors.

Apocalyptic digital spaces do not come into existence because the website's name contains the term "Rapture" or "End-Times." As discussed in Chapter 2, truths are not simply inscribed in place by default, but they emerge through social negotiations and discursive practices, and they need to be constantly reproduced to sustain their authority, validity, and legitimacy (Shapin, 1994; Foucault, 1980). Through the dissemination of evangelical apocalyptic conspiracist discourses on climate change, authors constantly validate and legitimize historically established truth-claims; they continue cultural traditions and make these digital spaces what they are: religious environments in which premillennial counter-knowledge can flourish.

Section III: The Revelation of Propaganda and the Suppression

Apocalypticism and conspiracism are discursive arenas which attempt to uncover hidden truths that are suppressed by the institutions in control of shaping public knowledge, such as universities, governments, and mainstream media outlets. Emphasizing the importance of the original Greek meaning of apocalypse as revelation or unveiling, DiTommaso (2020: 318) writes that the "primary operational function of apocalyptic speculation is to reveal the true nature of things to members of the group for which it is intended." Likewise, it is a core purpose of conspiracism to identify perceived lies spread by powerful cabals that govern human history and to establish alternative knowledge-claims as the truth (Lewandowsky, 2021; Uscinski and Parent, 2014; Barkun, 2013). Still, conspiracist End-Times discourses do more than attempt to reveal the hidden truths themselves. Rather, they also uncover the practices of knowledge suppression and methods used by the evil cabal to conceal the real purpose of, in this case, climate science and environmental politics. A key component of evangelical apocalyptic conspiracist ways of thinking is to ask how perceived false knowledge on climate change becomes established as true and how the supposed true knowledge on climate change becomes suppressed.

In that context, we argue that the apocalyptic conspiracist engagement with the practices that distribute perceived false knowledge and suppress true knowledge reinforces preexisting counter-hegemonic dispositions of apocalyptic evangelicalism. In short, evangelical apocalyptic conspiracists attempt to show that their knowledge is suppressed through the discursive practices of powerful media and political institutions. Misinformation campaigns concerning climate change are used to argue that the mainstream media and other epistemic institutions in power are both corrupt and an instrument of the apocalyptic evil. This claim of propaganda and an intentional suppression of perceived true knowledge eventually legitimizes evangelical apocalyptic conspiracism for adherents because there would not be a need for the suppression of knowledge if this knowledge would not be true and dangerous for the evil cabals in power. This conspiracist logic, which assumes that "information that appears to put a conspiracy theory in doubt must have been planted by the conspirators themselves in order to mislead," makes the conspiracist discourse non-falsifiable (Barkun, 2013: 7). In the digital evangelical apocalyptic conspiracist sphere, the revelation of propaganda and knowledge suppression is addressed through different discourses.

As a requirement for the revelation of propaganda practices, evangelical apocalyptic conspiracist knowledge must be established as the truth, whereas the dominant knowledge-power system of anthropogenic climate change must be established as a lie or hoax. Below, we first describe in the next section some discursive practices through which apocalyptic conspiracist authors define what is "true" and what is "false." Subsequently, we engage with the practices through which evangelical apocalyptic conspiracist authors portray themselves as epistemologically superior critics who have the capability to identify the malicious practices underlying the "regime of truth" of anthropogenic climate change. Seemingly immune to the Antichrist's deceptions, evangelical apocalyptic conspiracists discuss how anthropogenic climate change is established as the dominant knowledge through perceived propaganda practices. Through engagement with the practices of knowledge suppression, evangelical apocalyptic conspiracists portray themselves as victims of climate science, climate politics, and anti-Christian dictatorship. Since apocalypticism provides "an explanation for evil in the world and for the suffering of God's people" (Douglas, 2021: 2), it is not only argued that false knowledge is rendered true by the climate cabal but also that knowing Christians in particular are targeted by anti-Christian forces.

Revealing the Lies and the Truth

For any deconstruction of propaganda practices, the evangelical apocalyptic conspiracist knowledge-claim, which assumes that climate change is part of an anti-Christian superconspiracy, needs to be established as the audience's accepted belief and, consequently, true knowledge within the field of evangelical apocalyptic conspiracism. Certainly, it is a requirement for the conspiracist deconstruction of the dominant knowledge-power system of anthropogenic climate change that this dominant knowledge is infelicitous, whereas the apocalyptic conspiracist counter-knowledge is felicitous. The revelation of the hoax happens through a variety of practices addressed in this book, including that of biblical and scientific epistemic authority (see Chapter 6), appealing to preexisting apocalyptic geopolitical dispositions (see Chapter 7), but also simply through the discursive practice of denoting something as a "lie" or the "truth." Several authors use terms like "lies" or "hoax" when writing/talking about the dominant knowledge of anthropogenic climate change and "truth" or "facts" when presenting evangelical apocalyptic counter-knowledge. Evangelical apocalyptic conspiracists provide what they believe to be exclusive access to the truth about climate change and climate politics, which they cannot find in the "corrupt" mainstream media.

To name a few examples, EndTimeMinistries' Dave Robbins (2019a) promises on his podcast that "we will discuss the truth behind the [climate change] hoax," implying that he is going to "unveil" hidden and true knowledge. The Lamb & Lion Ministries (2020) call global warming a "political ploy which liberals are using in an attempt to gain greater control over society" in their online periodical Lamplighter. In a RaptureReady prophecy review from October 2021 (RaptureReady, 2021), climate change is denoted as a "SERIOUSLY STUPID LIE BY SINISTER GLOBALISTS" (capitals original). Also on RaptureReady, Davidson (2020) discusses the "whole Truth—Nothing but the Truth—it's GOD's WORD about Climate Change" in an article which claims that only God can precipitate a climate change during the End-Times:

> Have you noticed lately that all you hear from the 2020 Democratic candidates and what appears to be their ONLY "talking point" or "campaign promise" is CLIMATE CHANGE! None of these candidates probably even realize that, "YES," there will be a climate change in the soon-to-be future, but it is NOT the kind they are publishing as truth! In fact, they don't even know THE TRUTH—His Name is JESUS! Even the talking points about Climate Change they are publishing have been reported to be a lie from most modern scientists!

Alone, the discursive practice of denoting anthropogenic climate change as a "lie," "ploy," or "hoax" while apocalyptic conspiracist beliefs are described as the "truth" contributes to delegitimizing mainstream climate science and climate protection politics. We can apply Shapin's (1994) argument that as a discursive practice, terms like "knowledge," "truth," or "facts" constitute a judgment that contributes to establishing knowledge hierarchies in favor of the communicator. Relating to climate change, Djupe and Gwiasda (2010) explain that American evangelicals are more likely to trust climate change knowledge that is communicated by a perceived religious authority, in contrast to climate scientists who are often portrayed as outsiders. That is, because individuals are more likely to trust the knowledge-claims or judgments distributed by members of their social bonding group, group identity is an important factor in the subjective assignment of epistemic authority, competence, and credibility (Fricker, 1998, 2007; Goldman, 2002). Therefore, the sole practice of designating something "true" or "false" can change the audience's perception of climate change due to the *trust* in what they perceive as epistemological authorities, regardless of any derivation of scientific or religious knowledge-claims on climate change.

As in the above quotes, a reference to liberals or globalists further delegitimizes anthropogenic climate change because it is framed as a concern for political ideologies that are typically rejected in white and conservative evangelical apocalypticism. But more importantly, "the truth" is Jesus. The differentiation as to what is the truth and what is a lie is frequently accompanied by a powerful apocalyptic moral imperative that, for many adherents of contemporary evangelical premillennialism, dictates which side to take in the climate change debate. The belief in anthropogenic climate change is framed as supporting Satan, whereas the evangelical apocalyptic conspiracist knowledge-claims imply an alignment with Jesus and God. For example, on an episode of *The Hal Lindsey Report*, the show's host (Lindsey, 2015a) suggests climate change as "Satan's lie" as he calls for his adherents to put their "trust into God's word not in the self-serving false declarations of those who want to use this manufactured deception to bring people and nations under the control of just a very few" (2015a).

In the tradition of the Manichean dualism, only two options seem possible: one is either on the good side and rejects any knowledge of anthropogenic climate change or falls for the alleged climate change deception of the apocalyptic evil, because "one cannot "agree to disagree" when eternal life (or equivalent) is at stake" (DiTommaso, 2020: 318). Therefore, within the field of American evangelical apocalypticism, many "authors understand reality as a battlefield between opposing, dualistic spiritual forces—God/Satan, Good/Evil, Truth/Lies" (O'Donnell, 2020a: 2). In other words, a moral evaluation is connected to epistemic judgment about the truthfulness of knowledge-claims.

These binary moral judgments are further tied to evangelical apocalyptic group identity. Apocalyptic conspiracist discourses about climate change reproduce an "us-versus-them" geographical narrative, which is a fundamental and historically established characteristic of American millennialism and conspiracism (Hofstadter, 1964; Werly, 1977). Social power relations and knowledge hierarchies define this "us-versus-them" narrative. The competing knowledge-claims concerning climate change are not discussed as a conflict between equals, rather the evangelical apocalyptic authors we engaged with portray anthropogenic climate change as the dominant knowledge-power system, which profoundly affects political measures from the global to the local.

Contrary to their self-perception and social positioning as persecuted and culturally suppressed, evangelical apocalyptic conspiracists construct themselves as epistemologically superior. They commonly believe that the truth is only reserved for them since their ability to acquire true knowledge "originates from the one true reality, the sole source of wisdom," which in the case of evangelicalism is the Bible (see Chapter 6; DiTommaso, 2020: 318). Moreover, the supposed social or cultural suppression of American evangelicalism is a deliberate argument made by apocalypticists, as it increases the emotive power of their discourses. Regardless of their increasing influence on American politics and public opinion, religious End-Times conspiracists employ climate change to portray themselves as a suppressed minority.

Evangelical apocalyptic conspiracists are an example of how "a religious group may self-consciously and deliberately cast itself as an out-group—whether it is objectively disenfranchised or not—and launch its own counter-epistemic claims against perceived hegemony in order to claim its own space, its own power" (Robertson, Asprem, and Dyrendal, 2018: 3). It is this deliberate casting as an out-group that empowers conspiracist thought because apocalypticism is resistance literature (Megoran, 2012). It is part of the appeal of apocalypticism, as well as of conspiracism, to claim to provide *exclusive* access to the truth. Barkun (2013: 24) writes that this "cultural rejection is a powerful force" because esoteric or forbidden knowledge is particularly appealing. The suppression also validates counter-knowledge because it is argued that "the belief must be true *because* it is stigmatized" (2013: 28).

Apocalyptic conspiracism argues that "the conspiracy is so powerful, it controls virtually all of the channels through which information is disseminated," and apocalypticism, as well as conspiracism, become more powerful when the belief in the oppression is constantly reinforced and renewed (2013: 7). Consequently, many apocalyptic conspiracist prophecy writers in the digital evangelical sphere do not only denounce climate change as a lie or hoax but also elaborate on the malicious practices that establish the climate change hoax as the dominant truth.

Revealing the Propaganda: A Conspiracist Engagement with the Regime of Truth

Aware of the dominance of the knowledge-power system of anthropogenic climate change, several evangelical apocalyptic authors discuss how such anti-Christian evil managed to deceive the world. Or, to use the language of social epistemology, evangelical apocalyptic conspiracists examine how the regime of truth of anthropogenic climate change becomes established. This happens, consistent with the later work of Foucault (1980) and the social constructionist claims reviewed in Chapter 2, through the repetition of truth-claims by epistemic institutions in power (e.g., universities, mainstream media outlets, governmental institutions) and through the use of knowledge-creating strategies with a high degree of epistemic authority (e.g., quantitative and computer-based methods like the IPCC's climate models).

Apocalyptic conspiracist authors usually employ a less elevated terminology and use rather judgmental terms to denote the repetition of discourses that render something true. Frequently, they use a term that suggests that certain knowledge-claims are deliberately established to control public opinion: propaganda. The important distinction here is that the term "propaganda" implies that knowledge-claims become intentionally distributed to shape public opinion, whereas social constructionist perspectives do not necessarily imply any willful misconduct. Moreover, the use of the term "propaganda" depends on perceptions of the truth that are subjective and social judgments. DiMaggio (2022: 2) argues in his academic engagement with right-wing conspiracism that conspiracy theorists employ propaganda practices as "factually unsubstantiated conspiratorial messages." Countering this, the conspiracists we engage with employ the term "propaganda" to denote what they perceive as "factually unsubstantiated" knowledge-claims. In other words, what counts as propaganda depends on the epistemological predispositions of the space in which propaganda allegations are made. Within the digital spaces constituted out of evangelical apocalyptic conspiracist websites, mainstream media outlets and scientific institutions are the ones who advance propaganda. Outside of these and other conspiracist spaces, individuals who advance conspiracist beliefs are seen as the ones who use propaganda. What is propaganda and what is not depends on space, as space matters in the categorization of true knowledge and counter-knowledge. Still, both conspiracists and social constructionists use the "weapon" of deconstructing the relationships between language/discourse, knowledge, and social power. As the influential social constructionist Bruno Latour (2004: 230) argues in that context, "Of course conspiracy theories are an absurd deformation of our own arguments, but, like weapons smuggled through a fuzzy border to the wrong party, these are our weapons nonetheless"!

It is important to address the use of this weapon in the field of evangelical apocalyptic conspiracism since the alleged, or attempted, deconstruction of the dominant knowledge-power system of climate change legitimizes and empowers apocalyptic conspiracist discourses. Here, the supposed revelation of propaganda practices contributes to the non-falsifiability of apocalyptic conspiracist discourses because any counter-evidence can be denounced as propaganda and part of the superconspiracy. Moreover, the revelation of the propaganda amplifies the overall authority of apocalyptic conspiracists, as they seemingly demonstrate to their audience that they have the knowledge and tools to see through the deception and uncover the methods of the superconspiracy.

In that context, authors from the field of evangelical apocalyptic conspiracism argue that the repetitive communication of anthropogenic climate change as a major global issue contributes to the deception of the public and the establishment of a narrative of catastrophic human-made climate change/global warming as the truth. For example, an original article on End-Times Headlines (Snyder, 2021) deals with the "Insidious Agenda Of The Global Elite" and states:

> There is a consensus among the global elite that climate change is the number one threat to our planet by a wide margin, and that humanity is doing far more to cause climate change than any other source. They endlessly push this narrative through the news and entertainment companies that they control, and most people have bought into their propaganda on at least some level.

In this excerpt, the author states that the global elite propagates the narrative of climate change through "the news and entertainment companies they control," thereby, arguing in a conspiracist manner that evil elites have the power to control the media and to govern which truth-claims become disseminated by the media. The above quote also illustrates that the global elites are perceived as being successful with their propaganda. In that context, Snyder (2021) further explains that the fertility rate in the United States declined since 2007 because "their [global elite's] news and entertainment companies started to encourage women to wait longer in life to have children and to have fewer children when they finally did decide to have families." Also addressing the role of the media in the repetitive distribution of climate change knowledge-claims, EndTimeMinistries' expert on climate change, Dave Robbins (2019b), ascertains that "every news station is saying that it [climate change] is real and that it is one of the existential problems of our time" and continues that "the global warming climate change hoax is propaganda by the United Nations and other globalists to scare the populace into complying to their edicts" (2019b). In a lengthy article on RaptureReady, "the songwriter/musician, poet and writer from southern Oregon" (FaithWriters, no date) Mark A. Becker (2021) discusses the relationship between climate change and prophecy and asks how mainstream climate science became and stayed dominant knowledge by stating that a continuous replication and communication of a lie alone has a convincing effect.

How soon until these "experts" become known for what they are—political hacks and useful idiots? As Nazi propagandist Joseph Goebbels famously said: "If you tell a lie big enough and keep repeating it, people will eventually come to believe it." Add to this statement, "loud enough," and you have accurately defined the globalist elite's propaganda efforts on all fronts.

This Nazi comparison is a powerful one. The reference to Goebbels and the Nazi regime implies not only the existence of a suppressive regime attempting to control the world through the repetition of lies, but in particular that this was a historical reality and therefore could become one again. Here, Becker (2021) identifies Goebbels' propaganda practice of repeating a discourse as fundamental for the establishment of the dominant knowledge-claim of anthropogenic climate change.

Overall, there are numerous examples of claims of propaganda (either using the term or describing the practice) among evangelical apocalyptic conspiracists. It is not only climate change but also other important issues like Covid-19 or the LGTBQ+ movement, where adherents of evangelical apocalyptic conspiracism see their truth suppressed by the anti-Christian global government. Still, in-depth engagements that explain the explicit propaganda practices in detail are scant compared to allegations of a climate change superconspiracy that uses propaganda techniques. We could only identify a few examples which address how the propaganda apparatus functions or how the false knowledge of anthropogenic climate change is established as the truth. In the Global Warming Timeline on RaptureReady (no date), the so-called "Climategate" scandal is described as an example of the "fraud by the G20 propaganda machine." Climategate describes the leak of about 1,000 emails in relation to University of East Anglia-based climate researchers that allegedly prove that climate scientists work inaccurately, or even fudge data, to exaggerate the issue of climate change to justify increased funding. Although independent investigations on the email leak showed that "no scientific misconduct occurred" (Bricker, 2013: 219), ACC-skeptics and conspiracists still portray several emails that were taken out of context as proof of a climate change conspiracy. The evangelical apocalyptic conspiracist website RaptureReady (no date) portrays the email leak not only as proof of a climate change conspiracy that benefits the scientists but for a global superconspiracy:

Leaked Emails Expose US–UK Science Fraud on Global Warming—MissionGalactic Freedom reports leaked emails between the University of East Anglia and their U.S. counterparts show fraud by the G20 propaganda machine. The leak of those emails blew away the G20 scam and proved that this was simply a measure to create stealth taxes on the ordinary people of this world and for the poorer countries to be bullied in taking western technology and huge debts via the World Bank and the IMF.

In this way, RaptureReady draws from non-apocalyptic climate change conspiracy theories and reframes Climategate as part of an apocalyptic conspiracist worldview where the alleged scientific misconduct not only indicates that scientists want to increase their funding but also implies that the supposed manipulation of climate change data by the University of East Anglia researchers was instructed by G20 states as part of a globalist propaganda machine. The email leak is also addressed in the YouTube show/podcast The Pastor's Perspective by the Texan pastor Andy Woods (Woods and McGowan, 2017), who claims that Climategate shows how climate science is intentionally faked because of a "political result that people want" (which he ascribed to Satan's power of deceiving people).

Also addressing scientific institutions, Hal Lindsey (2015a) says that "this story [climate change] is not about science. It is about politics and propaganda." Establishing that climate science is a tool to deceive the public, Lindsey further addresses some scientific institutions and says that "the things that NASA says and the way they say them reflects the techniques of propagandist not scientist" (2015a). Although a detailed explanation of the supposed NASA propaganda techniques is not provided by Lindsey, the argument made here and elsewhere is that scientific institutions create false knowledge for the purpose of evil global powers that eventually serve the Antichrist. It is part of the propaganda machine that knowledge-creating institutions like NASA or the University of East Anglia distribute scientific findings that benefit the anti-Christian conspirators.

In another example of an alleged propaganda technique, an EndTimeMinistries (2019a) article mentions indoctrination of young children at school ordered by the government, in this case in Italy. Here, EndTimeMinistries refer to a Breitbart article which claims that "Italy's Leftist Government" is going to indoctrinate schoolchildren and that this "environmental indoctrination will be introduced as part of the students' civics education as a sort of 'Trojan horse' that will eventually 'infiltrate' all courses." In another article on EndTime Headlines, the founder of the website, Ricky Scaparo (2021), explains that YouTube announced they would not pay out any money to channels that challenge the scientific consensus that current climate change is anthropogenic. While Scaparo does not explicitly argue that this is a propagandist practice, it indirectly supports the belief in corrupt mainstream media companies which attempt to cover up the truth because the article is posted in a digital space in which it is established knowledge that climate change is an anti-Christian superconspiracy. New meaning is assigned to the article, which claims that the distribution of knowledge-claims is regulated by companies like YouTube. Eventually, the article contributes to supporting the belief in intentional propaganda.

Another critical engagement with the practices that establish the regime of truth of human-made climate change is a discussion of the specific terms "global warming" and "climate change." It is a frequently made allegation within the field of evangelical apocalyptic conspiracism that, in the public, the term "global warming" got intentionally replaced by "climate change." Here, it is argued that "global warming" did not live up to the expectations of the so-called global elites in scaring the population into accepting a global dictatorship or that the elites/scientists/media failed to prove that the average temperature of the globe increased in recent years. The more flexible concept of "climate change" then, supposedly, allowed these global elites to create renewed hysteria anytime the climate is reported to have changed or is projected to change.

In an example of the conspiracist engagement with the climate change terminology, The Lamb & Lion Ministries (2018) argues that "even the advocates of Global Warming have recognized the absurdity of their claim that humans are responsible for the earth's warming. So, Global Warming has morphed into 'Climate Change'." Although this quote somewhat confusingly suggests that the difference between global warming and climate change is anthropogenic activity, the argument made here is that the terminology changed to keep communicating a non-existent issue. Contrasting such shallow allegations, RaptureReady's Becker (2021) provides a more detailed explanation of the discourses of "global warming" and "climate change." Here, he states that the global cooling theory in the 1970s was the first attempt at establishing a climate hoax:

> When it comes to the "man-made" climate hoax, there can be no denying its existence when one just looks back on the past century. When I was a kid in the 1970s, the scare-tactic claim of the day was that we were about to enter into a little Ice Age [Hyperlinks: Time Magazine: Another Ice Age? And Newsweek: The Cooling World].

> Well, that didn't work! So, they needed something else. Out pops Al Gore and his ironically titled An Inconvenient Truth, and they're off and running again—this time in the other direction with something called "Global Warming." This didn't work either as, laughably, almost every time the climate cabal got together to discuss how they could "combat the weather" (just the thought of this claim brings my mind to laughter), temperatures at their meeting sites were trending below average and, in some cases, bitterly cold. It sure seemed that God was having good chuckle at their expense! But, then, they had an epiphany: Let's call it "Climate Change,"

and we can have it both ways; we can claim catastrophe when it's cold and when it's hot!

This wasn't radical enough for them, though, as some of the people—those with half a brain, anyway—were still rejecting their ridiculous claims. They needed peer pressure to help achieve their diabolical aspirations along with fearful slogans. So, they came up with new terms such as "Climate Crisis," "Climate Emergency," and "Climate Chaos."

As part of a lengthy article outlining his perception of the anti-Christian climate change conspiracy, Becker (2021) claims that since the 1970s, the "climate cabal" has been utilizing climate-related issues as a scare tactic. After the announced global cooling did not happen, it was apparently Al Gore who popularized the term "global warming." But since no global warming occurred, at least according to Becker, the term "climate change" was inserted into the lexicon by the evil elites to create a new concept for the crisis. Now, according to the conspiracist logic, any change in the climate could be portrayed as a catastrophe, and climate change would not be in disagreement with the scientific consensus.

In addition, Becker (2021) further addresses the emergence of terms like "climate crises" or "climate emergency." Also mentioning the presence of these relatively new terms, EndTimeMinistries (2019b) posted an article on its website which writes of the Guardian's conceptual shift from "'climate change' for 'climate emergency, crisis or breakdown,' and 'global warming' will be substituted for 'global heating'." Although EndTimeMinistries does not provide a comment on the prophetic relevance of the Guardian's new practices concerning climate change terminology, the article appeals to previously established beliefs of this digital space regarding propaganda practices explained in other articles on their website. Eventually, such articles empower the belief in the Antichrist's world government.

However, the conspiracist focus on the terminology of climate change shows that the recent proclaiming of "climate emergencies" or "climate crises" reinforces conspiracist climate change beliefs. While "global warming" and "climate change" are mostly objective terms which simply describe an increase in global average temperature or a changing climate, a "climate emergency" implies a catastrophic scenario, a crisis, and, in particular, the need for human action. Addressing the political meaning of a "climate emergency," Mike Hulme (2019) explains that "emergencies promise the mass mobilization of a jurisdiction's full economic, social, and technical capacities to ward off an existential threat" and "that in the case of climate change, such declarations are driven by a heightened sense of urgency among an array of scientists, activists, journalists, and others about the need to arrest climate change within the next 10 years" (2019). In short, terms like "climate crisis" or "climate emergency" imply the need for profound political action, in contrast to the rather neutral scientific terms "climate change" or "global warming."

Due to this political meaning implied in the term "emergency," conspiracists interpret the proclamation of climate emergencies as an intentional act within the superconspiracy to urge the population and governments to fully commit to globalist climate protection policies. The emphasis on the mere term "emergency" illustrates that apocalyptic conspiracist authors are aware of the relationship between language and power as well as the political meanings conveyed by individual words. But in light of preexisting conspiracist dispositions, they perceive it as another attempt to deliberately deceive. It is the radical commitment of the world that makes a "climate emergency" compatible with an apocalyptic conspiracist worldview because the expectation of a profound political and cultural transformation aligns with the expected apocalypse. Furthermore, the attention to the crisis aspect of climate change as well as the proclamation of climate emergencies made by several governments indirectly reconfirms the conspiracist worldview: "If they try that hard, it must be a conspiracy."

Summarized, there are several texts in digital spaces in which evangelical apocalyptic conspiracists attempt to reveal the methods which establish the dominant power-knowledge system of anthropogenic climate change. Some argue straightforwardly that globalist elites utilize propaganda practices to establish a belief in anthropogenic climate change. It is usually claimed that media outlets repeatedly distribute news on climate change to govern public opinion on the issue. Some authors engage with scientific institutions that are allegedly instructed to propagate false knowledge on climate change. Any utterance on propaganda techniques must be seen in relation to other evangelical apocalyptic conspiracist discourses, which claim that climate change is the Antichrist's deception to establish a world government. Since "all texts . . . are in an intertextual relationship with other texts" (Paltridge, 2012: 11), a detailed explanation of propaganda practices is not always necessary, as the mere accusation of a deliberate deception of the public falls in line with other claims made within the field of evangelical apocalyptic conspiracism. The diverse individual arguments addressed in this and the subsequent chapters of this book (e.g., eschatology, science, geopolitics) mutually support each other and form a powerful, self-sustaining discursive field that has an impact on the environmental attitudes of American evangelical premillennialists as well as non-religious apocalyptic conspiracists.

Several evangelical apocalyptic conspiracist discourses claim that the truth-claim of anthropogenic climate change became artificially established by the repetitive communication of this knowledge as true. Although any social epistemological language is not employed by evangelical apocalyptic conspiracists, this deconstruction of a truth which becomes established through repetition of discourse reminds one of social constructionist arguments. Harambam and Aupers (2015: 466) address the conspiracist deconstruction of the truth and argue that "conspiracy theorists compete with (social) scientists in complex battles for epistemic authority in a broader field of knowledge contestation." Many conspiracists not only "openly resist the 'regime of truth'" but also critically engage with the discursive practices through which the regime of truth becomes established: they attempt to uncover how knowledge, language, and power are intertwined (Harambam and Aupers, 2015: 467; following Bratich, 2008 and Foucault, 1970). The emphasis on the change in terminology, from "global warming" to "climate emergency," shows that apocalyptic conspiracists are aware of the power of language. Nevertheless, some evangelical apocalyptic authors see the change in climate change discourse as another indicator that proves the existence of a superconspiracy. The proclamation of climate emergencies is perceived as an almost desperate attempt to eventually scare the population into accepting restrictive climate protection policies after other tactics failed. When critical discourse analysis (CDA) asks how discourses enact, reproduce, and legitimate power in social and political contexts, then conspiracist critical discourse analysis (CCDA) can do the same. For them, the term "climate emergency" is a piece of language intentionally designed to increase the political and social power of the evil climate cabal and suppressive globalist elites. Such practices illustrate that conspiracists "adopt the label of critical thinkers for themselves" as they attempt to attack the dominant epistemic strategies (climate science), the produced knowledge (anthropogenic climate change), but also the communication of knowledge (Muirhead and Rosenblum, 2019: 109). Furthering these critical conclusions, Latour's (2004: 229) asks the rhetorical question: "What's the real difference between conspiracists and a popularized version of social critique, inspired by, let's say, a sociologist as eminent as Pierre Bourdieu?" Foucault is also mentioned as another example, which suits the provocative comparison made by Latour. So what is the difference between social constructionists and conspiracists, who both ask how dominant knowledge, in our case, climate change knowledge, becomes established and reproduced through the use of language?

Latour (2004) observes that ACC-skeptics utilize social constructionist methods to delegitimize human-made climate change and further suggests that some social constructionist research could have empowered ACC-skepticism. It is argued that powerful scientific institutions distribute climate change knowledge and that the mainstream media, or political institutions, repeatedly reproduce this knowledge so that the majority of society accepts it as true. They further argue that a certain language, such as the use of the word emergency, legitimizes more power than other words. The important distinction, however, is that evangelical apocalyptic conspiracists are driven by the belief that all of these discursive practices were intentionally designed as a means to exercise power or the increase the power of the Antichrist.

The sole purpose of CCDA is to legitimize and reinforce the preexisting belief in the evilness of the IPCC, UN, mainstream media institutions, and globalist or liberal politics. In other words, a deconstruction of the dominant truth exists to confirm their respective suppressed truth since the interest in discursive practices is driven by apocalyptic conspiracist confirmation bias. Demeritt (2001: 310) states that "demystifying scientific knowledge and demonstrating the social relations its construction involves does not necessarily imply disbelief in either that knowledge or the phenomena it represents," but for apocalyptic conspiracists, deconstruction *does* imply a disbelief in this knowledge. Moreover, Latour (2004: 242) writes that "critique is . . . useless when it begins to use the results of one science uncritically." Evangelical apocalyptic conspiracists lack any critique on their own science, their own truth. They uncritically use the knowledge of their own apocalyptic conspiracy belief system to critically deconstruct concurring knowledge. Only perceived lies can be deconstructed, but the perceived truth cannot. This is what differentiates conspiracists from constructionists. Social constructionists attempt to deconstruct not supposed lies but rather what they perceive as true and ask how social power relations, spaces, and their preexisting cultural and social beliefs affect the production of knowledge. For social epistemology and social constructionism, "the question was never to get *away* from facts but *closer* to them" (Latour, 2004: 231, italics in original). Still, the engagement with propaganda techniques is just one element of the deconstruction of the dominant truth. Other important techniques will be addressed in the subsequent chapter, which deals with the particular counter-epistemic practices employed by evangelical apocalyptic conspiracists.

Conclusion: "Yesterday's prophecies. Today's headlines" . . . and Today's Technologies

The tagline of *The Hal Lindsey Report*, "Yesterday's prophecies. Today's headlines," is the guiding practice in the American evangelical apocalyptic digital world. We have shown in this chapter that preexisting prophetic beliefs, as well as historically shaped social and political attitudes (Yesterday's prophecies) within American evangelical apocalyptic conspiracism, influence the perception of contemporary events and developments that relate to climate change (Today's headlines). Climate change-related discourses are perceived through a preexisting "lens" of evangelical apocalyptic conspiracism while at the same time, this perception of climate change reconfirms previously established beliefs of American evangelical apocalyptic conspiracists.

The engagement with contemporary primary sources from digital evangelical apocalyptic conspiracist spaces and a comparison with the academic literature on the history of evangelical apocalypticism have shown that historically established concerns of the field of American evangelical apocalypticism, like anti-liberalism, anti-globalism, and skepticism/hostility toward the pope, are reproduced in digital spaces. By relating partially century-old prophetic truths to contemporary events like climate change, contemporary prophecy writers preserve the overall apocalyptic conspiracist world in new spaces that are enabled by digital technologies. The continuation of evangelical cultural histories in digital spaces highlights the fact that "the digital" transcends any binary conceptions because beliefs developed in the physical world influence how knowledge on the internet is created and consumed, while at the same time, the apocalyptic conspiracist knowledge-claims retrieved from digital spaces influence practices in the physical world, whether that be elections or the storming of governmental buildings. The engagement with evangelical apocalyptic discourses on climate change has further shown that the Christian apocalyptic worldview is highly sustainable in online environments because preexisting beliefs can constantly be confirmed through the publication of new articles, videos, or podcasts. It is important to understand that each publication of a text fortifies the cultural and social composition of the digital space in which the text is published. These authors assign a meaning and a culture to their particular (web)sites, which then affects the perception of texts on this website.

In that context, an important practice we observed is the reposting of discourse from another (web)site into evangelical apocalyptic conspiracist digital spaces to assign new meaning to this piece of discourse. As the cultural histories and social dispositions of a space affect the consumption of knowledge-entailing discourses, texts from non-evangelical spaces become prophetically relevant when they are consumed in evangelical apocalyptic spaces because this text then stands in an intertextual relationship with previous texts from this space. Illustrative of the relationship to the rather secular conspiracist right, several articles in evangelical apocalyptic digital spaces originate from authors that do not include any reference to evangelicalism in their texts. Yet, these texts usually do not contradict any preexisting religious attitudes of evangelical digital spaces while confirming the cultural and political attitudes of evangelical apocalypticism. Even texts from mainstream media outlets can reinforce preexisting apocalyptic conspiracist dispositions when they are transferred into other spaces. For example, websites like endtimeheadlines.org regularly take texts from non-apocalyptic conspiracist, and even mainstream news outlets and repost them in their own digital environment if they can confirm their preexisting prophetic worldview. News reports on global climate protection policies are often simply copied onto websites like endtimeheadlines.org without any comments or changes, but readers will likely put these texts into context with other texts that argue for an anti-Christian superconspiracy and the imminency of the End-Times.

Climate change and, in particular, the politics of climate change will likely continue to be perceived as one of the most important signs of the End-Times. Hal Lindsey (2022) speculates that "over the past two years, they have primarily used Covid-19. In 2022, they will continue to use fear as the primary driver of change. But as the year goes on, expect them to increasingly shift the focus of our fears more toward climate change." Seven years after his initial in-depth engagement with climate change in relation to the Paris Climate Agreement, Lindsey still holds that the politics of climate change will usher in the anti-Christian world government, the Rapture, and the End-Times—and as a result, fifty-two years after the publication of *Late Great Planet Earth*, Maranatha, "the Lord is coming soon."

6

Generation of Evangelical Apocalyptic Climate Counter-Knowledge

A necessary condition to make the cultural, political, prophetic, and anti-globalist arguments of evangelical End-Times conspiracism work is to demonstrate that anthropogenic climate change and global warming do not exist, or at least to show that they are not serious global crises that require pervasive political responses. Therefore, rather than simply stating that climate change is a "lie" or "hoax," many authors within evangelical apocalyptic conspiracist digital spaces attempt to delegitimize the widely accepted truth of anthropogenic climate change by presenting or generating climatic and environmental counter-knowledge. Evangelical prophecy authors refer to religious and non-religious sources and utilize different forms of perceived empirical evidence to disprove the popular, powerful, and largely accepted scientific knowledge of anthropogenic climate change as well as connected epistemic authorities and strategies. In that context, one of the most important sources of the truth on climate change and the natural environment, at least for many American evangelicals, is the Bible. It is this prominently placed book that carries several episodes of *The Hal Lindsey Report* on climate change.

Addressing the climate change superconspiracy with the Bible under his left hand, Lindsey increases his epistemic authority for his evangelical audience as he conveys that his utterances are not his subjective beliefs, but biblical truths. As "people of the Book." the majority of American evangelicals ascribe a high amount of epistemic authority to the Bible (Roberts, 2008: 33). Yet, this use of the Bible must be approached as a counter-epistemic knowledge-creating practice since global epistemic authorities on climate change, like the IPCC, create climatic and environmental knowledge through a synthesis of research that is based on quantitative methods and computer models, and not the

Bible or any other urtext. Within the cultural environment Lindsey operates in, epistemic strategies that are based on biblical authority are considered valid ways of creating knowledge on the climate and natural environment (Noll, 2002; Evans, 2011; Evans and Feng, 2013). As spaces, their cultural histories as well as political and social dispositions affect which epistemic strategies are considered as valid ways of creating climate and environmental knowledge, practices that are considered to be "counter-epistemic" in the context of global epistemic power relations can still be valid knowledge-creating strategies in other scales or spaces, like American evangelical apocalyptic digital spaces.

This chapter examines the counter-epistemic strategies that are portrayed as valid ways of creating knowledge about the environment and analyzes how evangelical End-Times authors create, present, and legitimize anthropogenic climate change skeptic (ACC-skeptic) knowledge on global warming, climate change, and environmental change. We critically ask *how* and *why* relevant authors employ certain sources and ways of creating knowledge about the climate and also address the justification of this knowledge. Following the social epistemological framework outlined in the second chapter of this book, we hold that externally acquired knowledge is justified when it is communicated by an individual who is trusted and when methods are used that are perceived as valid and authoritative epistemic strategies (Fricker, 2012, 1998; Shapin, 1998, 1994). In the first section of this chapter, we explore how premillennial authors utilize the authority of the Bible to argue that humanity cannot substantially alter parts of God's creation, like the climate system. In the second section of this chapter, we examine the use of external sources of knowledge in which authors cite preexisting religious and non-religious knowledge-entailing discourses that argue against the existence of anthropogenic climate change. The third section engages with epistemic strategies in which authors create their own form of empirical knowledge on climate change through the interpretation of preexisting data, through first-hand observations, or a combination of both.

Section I: Biblicism and Knowledge on the Climate and Natural Environment

In the Bebbington Quadrilateral (Bebbington, 1993), biblicism is described as one of the four main features of evangelicalism which separates evangelicalism from other forms of Christianity and mainline Protestantism. Biblicism means a particular regard for the Bible as the "final authority on all matters of faith and practice and that all theology should be consistent with it [the Bible]" (Steffaniak, 2020: 358). American evangelicals commonly consider "the Bible the very word of God," which is "true not just as religion, but also as fact" (Noll, 2002: 6, 143). In short, the Bible is a trustworthy source of knowledge within American evangelicalism.

Many faithful American evangelicals assume that science and the Bible produce the same truths or that true science can only confirm, but not contradict, biblical facts (Evans, 2011). Following the social epistemological perspective of this book, any designation of a fact is a subjective judgment that depends on the cultural and social dispositions of the individuals or social groups that define what is to be approached as a fact. Hence, possible "diverse interpretations of the Bible" cause a theological diversification of American evangelicalism that can also be observed in relation to climate and environmental attitudes (Roberts, 2008: 39). Although most groups claim to stick to the creed of biblicism and biblical inerrancy, there are different conceptions of God's word that have strong implications for the perception of scientific knowledge and issues like climate change (Evans and Feng, 2013; Roberts, 2008). While there are also critical and self-reflective streams within evangelicalism that stress that any reading of God's word is always accompanied by the practice of subjective interpretation (Steffaniak, 2020; Kovach, 2012), the evangelical apocalyptic conspiracist authors we engage with portray their respective interpretation of the Bible as an objective and undeniable biblical truth.

Genesis is a frequently mentioned book of the Bible that in their reading proves that global warming, environmental change, and changes in weather patterns cannot be caused by human activity but rather only by God. Hal Lindsey (2015a), for instance puts an emphasis on Gen. 8:21, in which God claims that he "will never again curse the ground for man's sake" (Gen. 8:21). Lindsey (2015a) understands this as God's promise "never to destroy the present earth because of man's sinfulness." Lindsey rejects any connection between human activities and the natural environment, indicating his disbelief in human-made climate change. Additionally, Lindsey further states that he does not believe in any form of contemporary climate change or global warming. Lindsey interprets the verses of Genesis as divine promise which ensures that the Earth's climate system will not change since "cold and heat, winter and summer, day and night" (Gen. 8:22) will remain as long as the Earth, as a complex divine creation, will exist. Likewise, Dave Robbins (2019b) also refers to Gen. 8:22 in one of the EndTimeMinistries' Soundcloud/YouTube shows on climate change, as he claims that "the Bible says: 'While the earth remaineth, seedtime and harvest, and cold and heat, summer and winter and day and night shall not cease'."

In a similar manner, Andy Woods from the Sugar Land Bible Church in Texas states on his YouTube channel that it is "shocking . . . how many Christians believe in man-made global warming" (Woods and McGowan, 2017). Here, the

Texan pastors refer to God's covenant with Noah in Gen. 8:22, which "deals with the whole issue of climate change" (2017). Like Lindsey and Robbins, Woods and his colleague McGowan see this verse as a divine insurance for the stability of the natural cycles which God designed. Climate change, for Wood's, turns into a "theological issue" because Christians who believe that the world "is so defectively manufactured" that "humans could mess it up" allegedly cannot believe in the word of God and his power (2017).

Overall, the portrayal of verses from the book of Genesis as biblical facts that disprove anthropogenic climate change is a common practice within apocalyptic conspiracist digital spaces that is also used within the wider cultural realm of American evangelical ACC-skepticism (Kearns, 2014, 2012; Roberts, 2008). Similar to non-apocalyptic conspiracist organizations like the Cornwall Alliance, premillennial conspiracy authors use strategies of the American evangelical "climate skeptic playbook" that refer to Genesis to argue for an anthropocentric theological perspective that not only assumes that the planet was created by God for humanity to exploit but also that only God can control the natural environment and the climate system (Kearns, 2012: 143). Many evangelical ACC-skeptics propagate a theological perspective that assumes human-nature dualism and a "dominion" approach to the environment, which interprets Gen. 1:26 and Gen. 1:28 as explaining that God created nature merely as a space for humanity as a resource for human needs.

Within evangelical apocalyptic conspiracist digital spaces, the truth-claim that only God can control the climate is not always supported by specific Bible citations. For instance, in a Lamb & Lion Ministries talk show on Christian voting recommendations before the 2020 American presidential election, the prophecy blogger Marsha Kuhnley (2020) states:

> A lot of people today think that essentially that the climate is out of control, and that humans have the ability to do something about it. Right? And the Bible teaches something quite different. We know that God created the world and everything in it, and that God is in perfect control of His creation.

Kuhnley uses the authority of the Bible to argue against the possibility of anthropogenic climate change without any reference to Genesis or another book of the Bible. Such an utterance that simply mentions the Bible as a source of knowledge without providing a specific verse can be a powerful intertextual practice, as "discourse produced in one context inevitably connects to discourse produced in other contexts" (Hodges, 2015: 42). In this case, the words "the Bible teaches" automatically refer to prior discourse that derives the non-existence of human-made climate change from the Scriptures. Kuhnley's mention of the Bible can be understood as an intertextual discursive practice that reinforces already established knowledge as one of the "subtle mechanisms" (Foucault, 1980: 102) that legitimizes the End-Times truth of a climate change superconspiracy. The claim that only God can control the climate and environment is already a widely accepted truth within the digital space of the Lamb & Lion Ministries website and the End-Times blogosphere more broadly.

Moreover, evangelical apocalyptic conspiracist authors use the authority of the Bible to argue that Christians who do not share their respective interpretation of the Bible (which is sold as an objective biblical truth by the respective authors) cannot be true believers. It is argued that evangelicals who believe in anthropogenic climate change reject God's word and do not trust the Bible, and as such, authors question the faith of their audience if they do not share their Bible interpretations. For example, this practice becomes obvious when Lindsey (2015a) asks his audience if they are going to "believe in the flawed theories of man which are motivated by all kinds of his own interests, or, in the clear covenant of God, who said, as long as this Earth remains, there will be no man-made climate change" to remind them of the biblical authority which should govern an evangelical's understanding of the world. It is the perceived "citationality" of the Bible or God's word that assigns authority and thus, power, to the evangelical ACC-skeptic thought (Sturm, 2010: 138, following Butler, 1990). These ACC-skeptic Bible interpretations constitute justified knowledge for apocalyptic evangelicals who assign a high amount of epistemic authority to the Bible and premillennial prophecy writers. For apocalyptic evangelicals, the presented source, method, and communicator of knowledge can be trusted.

Section II: External Sources of Knowledge

In addition to the use of Bible references to prove that anthropogenic climate change cannot exist, authors within evangelical apocalyptic conspiracist digital spaces refer to external accounts of knowledge. As addressed in the second chapter of this book, it is virtually impossible to acquire comprehensive knowledge about the climate through first-hand experiences since "climate science cuts against the gain of ordinary human experience" (Jasanoff, 2010: 237). Simon Dalby (2013: 43) explains that "the difficulty with carbon dioxide is that while it is ubiquitous, we exhale it all the time, it is odorless and invisible, and a matter for technical measurement, not a sensuous matter of direct human experience." Consequently, many evangelical ACC-skeptic prophecy writers acknowledge that they are "epistemically dependent" on external sources and cite prior knowledge about climate change (Fuller, 2002: 278).

We provide examples for different "identity prejudices" (Fricker, 2007; Goldman, 2002) toward external sources of climate knowledge that are used by authors to confirm their respective climate change beliefs: first, the use of a mainstream science source; second, a non-evangelical alt-right source; third, an ACC-skeptic evangelical source; and fourth, an evangelical apocalyptic source.

First, Hal Lindsey refers to selected scientific research from sources which do not belong to his bonding group but, on the contrary, to the dominant episteme of anthropogenic climate change. Lindsey (2015a) cites a study by the United States Geological Survey (USGS), which implies that the Antarctic ice sheets grew in 2014, to then argue that this study confirms God's words in Gen. 9:8-13.[1] These Bible verses explain that after Noah's flood, God promised that he would never again destroy the world by a flood. Lindsey concludes that this divine promise disproves any form of global warming because an increased global temperature would cause the melting of global ice sheets and, thereby, a destructive flood. And according to Lindsey, the partial growth of the Antarctic ice sheets in 2014 that was observed by the USGS confirms his interpretation of God's word in Gen. 9:8-13. This portrayal of science as seemingly confirming ACC-skeptic interpretations of the Bible is representative of climate change denialist thought within evangelical End-Times conspiracism. The source of the knowledge hampers the categorization of justified knowledge. Lindsey's knowledge-claim could be considered as justified/warranted if Lindsey would trust the USGS and its knowledge-creating methods as he did not measure the Antarctic ice sheets himself. Trust toward the USGS is a requirement for his scientific confirmation of God's covenant with Noah. Yet, the USGS belongs to the dominant episteme of mainstream climate and environmental science, a body of epistemic authorities that Lindsey perceives as being part of propaganda institutions. For example, the USGS (Foley et al., 2013) writes that "reduction in the area and volume of Earth's two polar ice sheets is intricately linked to changes in global climate and to the resulting rise in sea level." The USGS also uses the IPCC's data in its examination of the sea ice decline. It is then somewhat unlikely that Lindsey perceives the entire institution of the USGS as a trustworthy source on climate change. Lindsey cites the USGS selectively, albeit a likely negative identity prejudice toward the organization. In Lindsey's case, it seems like information which challenges one's predispositions "is discounted, while information that supports them is accepted uncritically," a practice common for conspiracy climate change discourses (Douglas and Sutton, 2015: 101). Rather than trusting the USGS as a valid source of knowledge concerning climate change, Lindsey only cites the research which can be reconciled with his ACC-skeptic interpretation of Gen. 9:8-13 while ignoring the large body of research by the USGS which provides evidence for anthropogenic climate change.

Second, the selective use of external non-religious sources that fall in line with ACC-skeptic beliefs can also be found on RaptureReady, even though a different external source of knowledge is used here. In an attempt to show that science disproves anthropogenic climate change, Geri Ungurean (2019) cites an academic article titled "No Experimental Evidence for the Significant Anthropogenic Climate Change" by Finnish researchers Jyrki Kauppinen and Pekka Malmi (2019). But rather than examining the original research article by Kauppinen and Malmi, Ungurean copies a text from the alt-right-libertarian website ZeroHedge (Durden, 2019) which summarizes the Finnish researcher's publication. As cited on RaptureReady and ZeroHedge, the Finnish researchers Kauppinen and Malmi (2019) argue that the IPCC models "fail to derive the influences of low cloud cover fraction on the global temperature" and conclude that "anthropogenic climate change does not exist in practice." These findings are, according to ZeroHedge, supported by researchers from Kobe University in Japan who identified an umbrella effect in their examination of the effect of galactic cosmic rays on the Earth's climate and cloud coverage.

The researchers from Kobe University claim that "this phenomenon [high cloud cover caused by galactic cosmic ray, causing an 'umbrella effect'] has never been considered in climate predictions due to the insufficient physical understanding of it" (Ueno et al., 2019). However, the study's Japanese authors never indicated that their findings can be used to argue against anthropogenic climate change, nor were they cited in the Finnish ACC-skeptic study. The causal link between the Finnish study and the Japanese study to argue that the "'umbrella effect'—an entirely natural occurrence—could be the prime driver of climate warming, and not man-made factors" (Durden, 2019) was manufactured solely by the alt-right website ZeroHedge. In so doing, ZeroHedge attempts to create a coherent argument in which two sources of scientific authority are used to prove that human-made climate change does not exist. Furthermore, the Finnish research article was not peer-reviewed or published in a scientific journal but only online as a short draft, which was heavily criticized for flawed methods and false reasoning by several senior climate scientists (ClimateFeedback, 2019). Ungurean (2019) uncritically used the article as a "proof that this whole 'Global Warming' mantra is NWO garbage" and to argue that "the bottom line is that the U.N. has attempted to use this outrageous lie to regulate the economies of the nations of the world," which aligns her argument with conspiracist beliefs. But as Shapin (1994) explains, trust legitimizes the adoption of external knowledge because it would be impossible to verify all prior accounts of knowledge. It is important to remind the reader that social bonding groups and trust are not only defined by religion but also by political ideologies (Suhay and Druckman, 2015; Goldman, 2002). Ungurean's reference to the libertarian alt-right website ZeroHedge could still be explained by a high amount of trust due to a positive identity prejudice (Fricker, 2007). Another important practice here worth mentioning is that a non-religious ACC-skeptic article with perceived scientific authority is transferred from a secular digital environment into an evangelical space. ZeroHedge's article reconfirms religious End-Times beliefs in an intertextual manner, as the pre-existing dispositions and beliefs of a space in which a discourse is consumed affect the meaning of a knowledge-entailing discourse (Livingstone, 2005b; Shapin, 1998).

Third, conspiracists in evangelical apocalyptic spaces also use perceived climate science expertise from the American evangelical milieu to confirm their respective ACC-skeptic claims. For instance, the Texan pastors Woods and McGowan (2017) draw from both non-religious and religious scientific authority. Woods explains self-reflectively that he is "not a scientist" and hence decides to quote people "that are scientists" as he refers to "31,000 credentialed scientists that have signed a petition challenging the very notion of anthropogenic global warming" on http://www.petitionproject.org (Petition Project, no date). The Petition Project is a frequently cited source in ACC-skepticism that conveys the existence of a scientific consensus against anthropogenic climate change.

Farmer and Cook (2013: 450) argue that The Petition Project is an example of faked climatic expertise, as "around 99.9% of the scientists are not climate scientists." Nonetheless, the high number of signees suggests to uncritical readers that there are few experts and a large number of individuals with scientific authority who argue against human-made climate change. Therefore, the petition is a highly effective argument in ACC-skepticism as it assigns increased legitimacy to the ACC-skeptic argument (Cook, 2017; van der Linden et al., 2017). Woods and McGowan (2017) mostly attempt to gain scientific authority by referring to climate skeptic sources and explaining that global warming occurred during the time of the Vikings, long before Woods got his SUV. Woods (2017) also confronts Climategate to show how climate science is intentionally faked because of a "political result that people want," which he ascribed to Satan's power of deceiving people. According to Woods, climate science is not settled, but "propaganda" (2017). Moreover, Andy Woods and Jim McGowan refer to Calvin E. Beisner and the Cornwall Alliance to further shore up their scientific and evangelical credibility. The Texan pastors create a powerful assemblage of authoritative knowledge as evangelical experts on climate change, and a large number of scientists seem to disprove anthropogenic climate change. It is in particular the knowledge-claims communicated or created by members of the same social bonding group are more likely to be accepted as true (Suhay and Druckman, 2015; Shapin, 1994). In that context, the increased *trust* which exists among members of a social bonding group is of particular relevance among religious groups (Goldman,

2002), and therefore the reference to the evangelical Cornwall Alliance serves to empower Wood and McGowan's (2017) argument. Yet, Beisner likely does not endorse the premillennial eschatology of Wood and McGowan's Sugar Land Bible Church. The social bonding group and positive identity prejudice here are not necessarily defined by shared evangelical millennial doctrine but by the evangelical rejection of mainstream climate knowledge and a shared *Feindbild* of mainstream climate science, environmentalism, and proponents of climate protection policies more broadly.

Fourth, premillennial authors portray other premillennial prophecy writers as valid and authoritative sources on climate change. An example can be found in a 2020 issue of the evangelical prophecy magazine the *Lamplighter* (Lamb & Lion Ministries, 2018), which emphasizes the prophetic relevance of natural disasters and global warming.

> "Global Warming." This lead to a frequently asked question: "What is the prophetic significance of 'Global Warming?'" The answer is that the whole concept is a political ploy which liberals are using in an attempt to gain greater control over society. The fact that it is a non-issue was pointed out brilliantly in a 2020 article by Jack Kinsella, a writer for Hal Lindsey. He observed that 96% of all greenhouse gases and other co-called global warming emissions originate naturally from the oceans that cover five-sixths of the earth's surface. Thus, as he put it, "That leaves only 4% to divide up between human activity; volcanoes, rotting vegetables, cow flatulence, etc." It is estimated that the human portion of that 4% is 14%, and 14% of 4% is .056!

The authors of the magazine refer to the expertise of Jack Kinsella as a credible source of knowledge on climate change. Interestingly, they misquote Kinsella as the *Lamplighter* lists an article from 2020 as the source of their ACC-skeptic knowledge-claims. The quote in the text above is from an article titled "Confessions of a Flat Earther" (Kinsella and Garcia, 2010), which first appeared on Kinsella's blog omegaletter.com. The article's title is a sarcastic reference to a statement by the climate activist Al Gore, who compared climate change deniers to flat earthers. Yet, more important than correct citation is the name of Jack Kinsella, who died in 2013. Kinsella's scientific authority is constructed in particular through his social identity and connection to the evangelical elite, namely Hal Lindsey. As the former ghostwriter of Lindsey's TV shows and writer for the prophetic and highly ACC-skeptic blog OmegaLetter, Kinsella's name has a positive identity prejudice within the evangelical apocalyptic community. In the climate change debate, "beliefs about communicators are in many ways inseparable from an assessment of the evidence itself" (Lewandowsky, 2016). A reference and association to Lindsey, whom many evangelicals still perceive as a valid source of knowledge, therefore assigns the ACC-skeptic knowledge-claims from Kinsella in the *Lamplighter* increased trustworthiness and legitimacy (Hahn, Harris, and Corner, 2016: 192).

Section III: Internal Knowledge, Personal Interpretations, and Observations

In addition to the use of the Bible and references to external sources of knowledge, some writers we engage with personally generate ACC-skeptic knowledge, either through the interpretation of existing temperature/climate data or personal first-hand observations. In both cases, the emphasis on data reflects the evangelical premillennial preference for the inductive Baconian scientific method, which seeks to generate knowledge through the interpretation of empirically observable data rather than the generation of predictive climate data through post-normal science (Roberts, 2008; Marsden, 1980). The Baconian method can be described as an inductive approach to science in which observations in the real world are used to make generalizations or generate a new theory, while generalization which exceed the observed facts should be avoided. Evangelical End-Times writers do not reject science but personally employ scientific practices and play the "game of science" to delegitimize the mainstream scientific consensus on climate change and establish alternative knowledge-claims (De Pryck and Gemenne, 2017: 123).

An important practice that can be observed in digital evangelical premillennial spaces in the interpretation of existing data is the ACC-skeptic epistemic strategy, which Farmer and Cook (2013: 451) call "cherry-picking." Cherry-picking describes a "focus on select pieces of data, often out of context, whilst excluding any data that conflicts with the desired conclusion" (2013). Similarly, premillennial authors attempt to disprove human-made climate change through cherry-picked observations that weather patterns did not change significantly in recent decades. This happens through isolated examples while "ignoring others that lead to the opposite result" (2013). Following that practice, RaptureReady's Sally Law (2018) claims that "the hottest day on record is not recent. Climate change proponents would like you to ignore the hard facts." The "hard facts" provided by Law are four single temperature records, which show that the hottest temperature ever recorded was in 1913 in California, while the lowest temperature recorded was in Eastern Antarctica in 2010.

These single temperature records give the impression that there is no global warming, as the hottest day ever was recorded more than 100 years ago, whereas the coldest day happened recently in 2010. Law comments that there were "not many cars on the road at that time," implying that there is no relationship between fossil fuel emissions and hot temperatures. Such individual temperature records from different places are not appropriate to argue for or against global warming as they are isolated, "cherry-picked," local examples which cannot reflect long-term global temperature trends. Without further explaining the meaning of these temperature records, Law (2018) argues that "radical Climate change solutions are theoretical, not something proven to make a significant difference in our weather patterns worldwide" and, thereby, manufacturer uncertainty (Hansson, 2017). The practice of presenting temperature data together with a "source" and the use of the word "fact" gives the impression of scientificity and empirical truth. For uncritical readers, the mere existence of quantitative data can have persuasive effects.

Irvin Baxter (EndTimeInc, 2017) from EndTimeMinistries employs a similar strategy of presenting isolated, local weather data to disprove global warming. Baxter does not draw from externally acquired weather data but provides his personal experiences and first-hand observations. In that context, Shapin (1988: 375) writes that "within empiricist schemes of knowledge the ultimate warrant for a claim to knowledge is an act of witnessing," which makes personal observations a powerful source of knowledge. Baxter (EndTimeInc, 2017) begins his podcast on "The Global Warming Religion" by stating that he personally experienced that "temperatures were 100 degrees above 39 days in a row" when he moved to Dallas 11 years ago, but he does not "think temperatures reach 100 degrees more than once or twice this last year." The anecdotal temperature data of two isolated years in Dallas cannot be used to prove or disprove climate change, but "there is much evidence that people's acceptance of climate change is a function of perceived temperature on the day" (Lewandowsky, 2021: 5). Therefore, these subjective experiences can affect climate change attitudes (Cook, 2020, 2017).

Such first-hand cognitive weather-related experiences and anecdotal evidence are powerful resources which can affect an individual's perception of climate change in both ways, increasing the acceptance of human-made climate change but also ACC-skepticism (Bergquist and Warshaw, 2019). Another example is found in the article "Global Warming" by RaptureReady's Steve Schmutzer (2017). Schmutzer observed, for instance, changeable weather patterns at his home in Colorado:

> In the past two weeks here in remote northern Colorado where I live, we've had two feet of snow, sunny days, subzero temperatures, 50 mile-an-hour winds, torrential rains, oppressive heat, blue skies, gray skies, and orange skies. There have been fogs so thick I couldn't see the sky at all. The climate has changed around here every couple of days.

Through a brief report of his personal observations, Schmutzer attempts not only to demonstrate that what he considers to be the "climate" changes within a couple of days, but also that this has been the case for centuries as he compares his observations to personally acquired descriptions of weather patterns from the nineteenth century. Schmutzer further states:

> It's been like this for the last hundred-plus years. We have an old alabaster mine on our property, and I've read diaries of folks from the 1800s that used to wrest the prized white rock from the ground. They described weather patterns like those now. To quote an ancient climate-themed truth, "...there is nothing new under the sun" (Eccl. 1:9).

Schmutzer's observations of isolated, local weather patterns in Oklahoma do not match the scale of *global* warming or *climate* change, although the Bible quote, which he describes as an "ancient climate-themed truth," potentially assigns further authority to his argument. Certainly, the use of isolated, small-scale, and anecdotal data is not exclusive to the evangelical End-Times conspiracism (Lewandowsky, 2021; Cook, 2020, 2017). The rejection of post-normal science connected to a "naïve appeal to Baconian empirical thought" (Roberts, 2008: 204) among conservative and dispensational evangelicals (Sutton, 2017; Noll, 2002; Marsden, 1980) can assign epistemic authority to isolated and sometimes personally experienceable temperatures as observable facts. Nonetheless, this must be interpreted in light of the political discourse surrounding the climate change debate as conservative evangelicals are more likely to oppose scientific findings which imply political action (Kearns, 2012; Evans, 2011; Noll, 2002). Here, the evangelical conspiracist way of knowing about climate change exists only in counter-dualistic reference to the IPCC and mainstream climate science. But as illustrated in Chapter 4, the EEN also uses observable climate data from the past and not the post-normal IPCC science, somewhat in the tradition of the Baconian approach. The EEN's authors also provide anecdotal evidence about experienced weather events that they employ to argue for anthropogenic climate change or global warming. For instance, the EEN often explains that people should just use their "Doppler Radar," meaning windows, to observe that long-term weather patterns, or the climate, have changed in recent decades (Hescox and Douglas, 2017). Therefore, the preference for different scientific methods alone cannot explain the rejection of anthropogenic climate change knowledge within evangelical apocalypticism.

Still, some authors discursively construct a contradiction between two different ways of knowing. The contradiction between two scientific approaches concerning climate change might be best illustrated by RaptureReady's Dan Payne (2016), who provides what he considers a legitimate scientific discourse. Summarized, Payne argues, based on personal calculations, that the sun is the main driver of climate change. With the intention of disproving anthropogenic global warming, Payne addresses his article to the climate change activist and actor Leonardo DiCaprio. Payne claims that his article is real science, which produces empirical truths as he writes: "Mr. DiCaprio, this article will now boldly dive into 'science' 'facts' and 'empirical truths'—if you happen to be reading, which is highly unlikely but, here goes anyway . . . ." The use of the terms "science," "facts," or "empirical truths" is a direct response to a statement by DiCaprio in which the actor argues that people who do not believe in anthropogenic climate change should not be allowed to hold public office. Thereby, Payne's article highlights what is important out of a social epistemological perspective: different perceptions of what constitutes a "fact" regarding climate change as well as distinct imaginations of legitimate science within American society. While anthropogenic climate change is a widely accepted truth for many individuals, such as DiCaprio, it is considered a false claim within the digital apocalyptic space of RaptureReady.

Payne's (2016) empirical truth-claims state that the sun is the primary driver of temperature changes on Earth as he argues that "a simple examination of the sun's impact during seasonal changes provides significant evidence of the power it has over the earth's climate." In that context, Payne (2016) engages with "the tilt of earth's axis" and calculates that "the total distance the North Pole fluctuates to and from the sun during the course of a year is only 0.0086 percent of the total distance that exists between the earth and the sun." Here, Payne observes that this relatively small change in distance can cause "as much as a 100 degree difference in temperature," which is "based solely upon the temperate zone during the same season in the same hemisphere." Eventually, Payne (2016) concludes:

> The power of the sun has been affecting the temperatures on earth tens of degrees every season and every year for thousands of years. Knowing this, isn't it far more plausible that an increase in solar activity can have a far greater impact on earth's weather than carbon emissions or bovine flatology?

To give further authority to that knowledge-claim, Payne cites two external sources of knowledge intended to provide further legitimacy to his own calculations and conclusion. First, Payne cites a statement by the Russian researcher Dr. Helen Popova, which was published in the British amateur astronomy magazine *Astronomy Now*. Popova states, based on historic changes in the sun's activity and naturally occurring changes in global climate, that "there is no strong evidence, that global warming is caused by human activity. The study of deuterium in the Antarctic showed that there were five global warmings and four Ice Ages for the past 400 thousand years." Although Payne acknowledges that "believers in biblical creation take issue with the timeline that Dr Popova presents," he uses her perceived authority as a "Dr." to argue that the sun's declining activity might cause a new Ice Age. Although Popava's statement contradicts Young Earth creationism, which is a popular belief among conservative evangelicals (Roberts, 2008), Payne uses it to achieve his objective of disproving anthropogenic global warming. The second source, somewhat surprisingly, is NASA, an institution which generates and supports knowledge of anthropogenic climate change. Payne (2016) writes:

> Another interesting twist on the power of the sun can be found in a study published by NASA scientists from the Goddard Institute for Space Studies in August 2016. The study suggested that the planet Venus may have been habitable in ancient times. The NASA scientists used the same computer models that they use to predict future climate change on earth. The most interesting part of the article is that the scientists believe that the ancient sun that was up to 30 percent dimmer than it is today. Think about that, their entire study was based upon the fact that a less powerful sun of the past could have allowed for the planet Venus to support life. So then it stands to reason that they must believe that the sun is now growing in power as time goes on. Astounding . . .

Aware of NASA's general position on climate change, Payne (2016) commits a fallacy of incomplete evidence. He uses one conclusion out of a scientific study, which, considered in its entirety, would lead to a different conclusion. Similar to examples discussed in the second section of this chapter, the ACC-skepticism that can be supported through external sources somewhat exceeds potential negative identity prejudices toward certain sources of knowledge that belong to the mainstream episteme of anthropogenic climate change, whether NASA or Popava's research that contradicts some evangelical beliefs in Creationism. But since "any topic can be framed in exactly the way a communicator desires if it is not presented objectively, honestly, and with context" (Harvey et al., 2018: 282) within the field of ACC-skepticism, Payne is able to construct a powerful knowledge-claim through his combination of personally generated knowledge and external sources of knowledge.

Conclusion: Motivated Reasoning?

This analysis of digital discourses illustrates that several knowledge-claims can be considered as justified whereas some forms of climate counter-knowledge within evangelical End-Times conspiracism lack justifications while still others contain logical fallacies. It is not about the truth itself but rather about the methods that are employed to generate or acquire a certain knowledge about climate change and global warming.

Certainly, the appeal to biblicism described in the first section of this chapter must be acknowledged as being a justified counter-epistemic strategy that has a high amount of authority within American evangelicalism. As space matters concerning the meaning and power of a knowledge-claim, it is important to acknowledge that Bible references are powerful practices within American evangelical spaces that increase trust toward a communicator of knowledge. Still, authors also use a variety of external sources to validate their ACC-skeptic attitudes. Here, we identified different levels of identity prejudices or trust toward the external communicators of climate knowledge. Shapin (1994: 9) argues that acquired knowledge can be considered justified/warranted if an individual trusts an external source to be a "reputable and veracious" source of knowledge. In several of the above discussed examples, trust can be explained through positive political, evangelical, or premillennial identity prejudice, generating knowledge which can be described as justified in light of the cultural, political, and epistemic beliefs of American evangelical apocalypticism.

The addressed references to the NASA or the USGS contradict any assumption of trustworthiness and, consequently, justified knowledge. We argue that the frequently cited practice of motivated reasoning can also be identified within premillennial conspiracism. Following the ACC-skeptic practice of weighing the epistemic authority of one article which supports the preferred narrative higher than a vast amount of opposing research, premillennial authors use sources (e.g., NASA, USGS) that are believed to be part of the climate change propaganda to legitimize an ACC-skeptic argument. Moreover, the preference of inductive Baconian science might persuade readers to reject the post-normal IPCC science, but it cannot explain the logical fallacies that emerge when local temperature data are applied to the large-scale, global phenomena of *global* warming or *climate* change.

7

Climate Change and Evangelical Apocalyptic Geopolitics

Religious apocalypticism constructs and affects imaginations of global space and international power relations, as well as the perception of issues of international importance, such as climate change and environmental geopolitics more broadly (Sutton, 2017; Sturm, 2010; Dittmer and Sturm, 2010). The environment is an "outsider," a threat to humankind, while the premillennialist is the "insider," the one with the knowledge to break free of the shackles of global tragedy by accepting God's great plan for the end of the world. This is a very political move, one that excludes "us" from "them," "inside" from "outside" through the border between Heaven and Environment.

This chapter deals with the interdependencies between apocalypticism, geopolitics, and climate change and analyzes how evangelical apocalypticists utilize climate change and related discourses to produce, or reconfirm, American anti-globalist geopolitical imaginations, and conversely, how climate change is used to demonize political actors and institutions as satanic globalists. While the employed terminology varies (e.g., one-world government, global government, New World Order), it is widely assumed within contemporary American evangelical apocalypticism that a form of suppressive global governance will usher in the biblical End-Times (Wood and Douglas, 2018; Barkun, 2010). Within this field of signs, evangelical apocalypticists argue that neither climate change nor global warming are real issues but rather faked crises which are used to gain acceptance for restrictive climate protection measures on a global scale, which then establish a system of control and surveillance under the pretext of climate and environmental protection. This deception is believed to be part of the Antichrist's plan to control the world, as allegedly prophesied in the Bible. In that context, foreign and intergovernmental institutions like the European Union or the United Nations, as established *Feindbilder* of the field of evangelical conspiracism, are often accused of conspiring against the

United States and American Christians by advancing the establishment of a dictatorial "Antichrist system" (Barkun, 2010: 122).

Concerning imaginations of global space and world politics, we argue that American evangelical premillennial discourses on climate change that assume a superconspiracy serve three interrelated geopolitical purposes. In what follows, the first section argues that climate change and related discourses are used to reinforce preexisting evangelical anti-globalist dispositions, to delegitimize intergovernmental and foreign institutions, and, generally, to reinforce the "anti-globalization bias of contemporary millennialism" (Barkun, 2010: 127). We explain in the second section how climate change and environmental discourses are utilized to empower American exceptionalism by claiming that the United States must constitute a sovereign and isolationist holdout to satanic and globalist endeavors through resisting global climate protection policies (O'Donnell, 2020a, 2020b). This fusion of American exceptionalism, sovereign protectionism, and religious End-Times discourse in which America becomes emplotted in premillennialist eschatology is part of American evangelicals' personal and collective identities (Dittmer, 2010). In particular, the presidency of Donald Trump and his America First politics that led to the withdrawal of the Paris Climate Agreement are put in context with the premillennial beliefs of his electoral base and their imagining of a special or elect role of the United States in the biblical End-Times script.

While the first two sections could suggest a rather simple Manichean imagination of global space in which the United States constitutes the "good" side, we illustrate in the third section of this chapter how evangelical apocalypticists construct a dualistic structure of good and bad, or insiders and outsiders, that does not necessarily emerge from a simple two-dimensional spatial imagination in relation to the border of the United States. Instead, the concept of a globalist, an American proponent of global governance, is used to define "evil others" (Sturm, 2010). Here, any endorsement of climate protection and environmentalism can be portrayed as advancing globalism. The *Feindbild* of a globalist can be applied to the Democratic Party, environmentalists, or entire states like California that promote climate protection policies, demonizing Americans as supporters of apocalyptic evil (O'Donnell, 2020a, 2021).

Section I: Anti-Globalism and the United Nations

A necessary disposition which makes apocalyptic conspiracist discourses on climate change work is to interpret, and believe that the Bible clearly prophesied the rise of a satanic global government prior to the apocalypse. In the American premillennial movement, the belief in the order of global politics before Christ's

Second Coming is commonly accepted and, thereby, a form of geopolitical knowledge, insofar as it imagines a global division of spaces. It is a prophetic truth for many American premillennialists that the Antichrist, demons, or an apocalyptic evil more generally will rise to power through globalist endeavors and intergovernmental as well as supranational organizations (O'Donnell, 2021, 2019; Sutton, 2017; Barkun, 2013). Still, this truth-claim, or knowledge, must be approached as "products of historically distinctive disciplines and forces" since it is not the Christian scriptures nor evangelical theology itself which inevitably dictates, such geopolitical imaginations (Asad, 1993: 129). Rather, contemporary evangelical apocalyptic anti-globalism develops historically and illustrates interdependencies between geopolitical, American nationalist, and religious discourse (Sturm, 2013; Agnew, 2006).

Evangelical apocalyptic anti-globalism is both a religious and a political attitude. An important religious legitimation for apocalyptic anti-globalism is found in Dan. 7:23-24.[1] Daniel's description of a kingdom which devours the whole Earth during the Tribulation, as well as other Scriptures which indicate the rise of a global evil or the decay of nation-states (for instance, Lk. 21:25[2]; Rev. 13:7[3]), led many American evangelicals to believe in the prophecy of an apocalyptic one-world government. During the second half of the nineteenth century, John Nelson Darby in particular popularized the idea of an Antichrist who gains dictatorial control of the world at the mid-time of the seven-year Tribulation period (Barkun, 2010). Nonetheless, it was not Darby alone who turned Bible prophecy into a source of geopolitical knowledge. Rather, a geopolitical reading of the Bible was repeatedly propagated by various influential American evangelicals in different social and historical contexts (Sutton, 2017; Gribben, 2011). For more than a century, powerful American evangelicals and Christian nationalists such as Hal Lindsey, Pat Robertson, or Tim LaHaye positioned nationalist threats to American freedom and sovereignty in context of the rise of a global evil as supposedly predicted by premillennial eschatology (Sturm and Albrecht, 2021b; Barkun, 2013, 2010; Gribben, 2004; Herman, 2001, 2000). During the twentieth century, and depending on the events of global politics, several international institutions or actors were portrayed as constituting, or cooperating with, apocalyptic evil. Through merging religious, conspiracist, and geopolitical imaginations, American evangelicals variously positioned the League of Nations (after First World War), the Soviet Union (during the Cold War), the European Economic Community/European Union (since the 1970s), and the United Nations not only as possible threats to American sovereignty and hegemonic power but also as the satanic powers which will subdue the world (Sutton, 2017; Gribben, 2011; Herman, 2001).

In particular, after the end of the Cold War, the idea of a global government, one-world government, or New World Order (NWO) became attractive monikers for premillennialists to explain geopolitics. While prophecy writers often emphasized wars as indicators of the End-Times, they now argued that global peace was a sign of the times (Barkun, 2010). Although this might sound contradictory, such a narrative flipping illustrates the important practice of adjusting prophecy to fit contemporary world politics. As Boyer (1992: 326) put it: "War fears or peace hopes—it made little difference to the skilled prophecy interpreter. All could be fitted into the end-time scenario." Regardless of the actual state of the world, repetitions of apocalyptic anti-globalist discourse in different historic contexts contributed to rendering this geopolitical knowledge as true, establishing an apocalyptic regime of truth. Based on the imagination of the Antichrist's "unprecedented political control," many American evangelical prophecy writers turned into "serious students of geopolitical developments" (Sutton, 2017: 19).

Using the Bible as a source of knowledge, the End-Times authors we engage with embed climate change within the historically established belief in the rise of an apocalyptic world government. One illustrative example is EndTimeMinistries' Dave Robbins (2019b), who provides an in-depth explanation of how the Bible prophesies a devilish, and climate-friendly, United Nations. Robbins cites several Bible verses and claims that the "prophesised world government" is described in Rev. 13:1-2 as a "beast" which got "power" and "great authority" assigned by the "dragon" (which is usually thought to be Satan). According to Robbins, the nations/kingdoms mentioned in Daniel 7 have "federalized into a one-world governing body" to form the beast described in Rev. 13:1-2. A "very key statement in Revelation 12:9" of "the Devil, and Satan, which deceiveth the whole world" leads Robbins to conclude that "this is exactly what the global warming climate change hoax is designed to do. The United Nations, the world government in the Earth today, is Satan's effort to establish his kingdom on Earth."

When listening to Robbin's podcast, it appears that evangelical anti-globalism and claims of a global climate conspiracy can be simply legitimized through the Bible. However, there is an intertextual gap between the cited words of the Bible and Robbins' (2019b) geopolitical interpretations. In this intertextual gap, new meanings are assigned to old texts even though the original texts are too generic or abstract to have actual implications for contemporary social contexts (Hodges, 2015). In Robbins' (2019b) case, this means that descriptions of a "beast with great authority" are applied to the United Nations, whereas climate change is believed to be the deception indicated in Rev. 12:9. Still, this gap is ignored to achieve the legitimization of the desired interpretation. Alone the discursive practice of reproducing biblical texts in the context of apocalyptic climate change conspiracism suggests that the Bible has actual implications for the geopolitics of climate change. Moreover, the above-mentioned apocalyptic anti-globalist regime of truth in the social environment of American (online) apocalypticism allows for authors like Robbins to create a connection between the Bible and global climate change policies regardless of intertextual gaps. His knowledge-claim is likely to be accepted due to the historically established apocalyptic geopolitical dispositions of his evangelical audience that are constantly reproduced in digital spaces through publications on climate change, Covid, the Great Reset, Agenda 21/30, or the NWO. Here, there is a reciprocal relationship between the preexisting geopolitical knowledge of American evangelical premillennialism and apocalyptic conspiracist climate change discourses. While premillennial conspiracists can draw from preexisting knowledge to legitimize their utterances on a supposed climate change deception, their texts, videos, or podcasts manifest the apocalyptic anti-globalist regime of truth.

But EndTimeMinistries' Dave Robbins is just one example of this practice. Several American evangelical prophecy authors draw from, and conversely reinforce, apocalyptic anti-globalist geopolitical knowledge despite any intertextual gaps. For instance, RaptureReady's Geri Ungurean (2016) also cites Revelation 12:9[4] to argue that the United Nations constitutes a satanic global government without an explanation of the apparent intertextual gap. Also on RaptureReady, Jonathan Brentner (2019), in his article on climate change and the NWO, claims without mentioning a specific Bible verse that "the book of Revelation tells us that, during the tribulation, the antichrist will exert worldwide control through a totalitarian world order."

Similarly, Hal Lindsey (2015a) argues that climate change is a deception to justify restrictive climate protection measures on a global scale, which eventually brings the Antichrist into power. To legitimize this interpretation of climate change, Lindsey refers to the books of Daniel and Revelation, which in Lindsey's premillennialist understanding, "both describe the Tribulation as a time of centralised governmental authority on the Earth" (Sturm and Albrecht, 2021b). In another example, RaptureReady's and RaptureForum's Daymond Duck (2020) refers to the "distress of nations" named in Lk. 21:25[5] to argue that "Climate Change is a hoax that globalists use as a scare tactic to get people to accept world government." Also on RaptureReady, Britt Gillette (2019) too argues that the "millennial" generation pushes certain issues, such as climate change, "at the forefront of the political agenda" to support "the globalist vision" which confirms the prophecy of "powerful global government" made in Dan. 7:23. There are several examples of this biblical citationality in apocalyptic discourses on climate change within American apocalyptic digital spaces. This citationality assigns authority and, hence, power to apocalyptic geopolitical discourses, as they refer to the Bible as the highest form of authority in American evangelicalism.

Nevertheless, there are also articles on climate change within religious apocalyptic digital spaces which do not refer to the Bible but still speak of the rise of a world government. Rather than employing biblical citationality or legitimizing their knowledge-claims with biblical authority, some authors simply state that climate change must be interpreted through a global apocalyptic conspiracy. On EndTimeHeadlines, Michael Snyder (2021) talks in a conspiracist manner about a "global elite," which "would like all of us to radically alter our behavior in order to combat climate change." In a similar manner, End-Times blogger Expat Gal (2020b) from Z3 News writes of "globalists" in a conspiracist article on an alleged climate lockdown without proving that the Bible predicted the rise of globalists or an Antichrist's NWO—although the Bible is later used in the same article to argue that anthropogenic climate change is a hoax.

Even though no direct Bible references are linked to the NWO in such articles, the spatial setting of the internet enables an intertextual relationship to prior discourse. A website or blog establishes a truth-claim over a longer period of time in numerous articles and prophetic contexts, so that the returning reader already "knows" about the relationship between geopolitics, the apocalypse, and assumed devilish globalist endeavors. This applies not only to single websites but also to the digital knowledge network constituted out of different American evangelical apocalyptic blogs and websites which share authors or frequently link to each other. The simple repetition of the truth-claim that a climate change deception indicates the rise of the Antichrist's global reign, or an apocalyptic evil, is enough to further reinforce the anti-globalist regime of truth within evangelical End-Times conspiracism.

Concerning the institutional dimension of conspiracist End-Times anti-globalism, the highest eschatological relevance today is ascribed to the United Nations (UN). Herman (2001: 58) states that the American Christian right provides "significant leadership to the anti-UN movement in the USA" not only because the American evangelical apocalyptic movement views the UN as a globalist agent of Satan (Dittmer, 2007) but also because of the UN's environmentalist commitment, which many evangelicals perceive as an anti-Christian worship of the Earth (Herman, 2001; Spark, 2001). Ever since Tim LaHaye's *Left Behind* novels, which emphasized the relationship between the UN and the Antichrist, the intergovernmental organization has remained at the center of American apocalyptic beliefs (Dittmer and Spears, 2009; Gribben, 2004). Research by Chaudoin, Smith, and Urpelainen (2014) even suggest that American evangelicals are more likely to oppose climate protection policies when the UN is involved, in contrast to domestic climate protection, as they are deeply skeptical of international cooperation. Consequently, many authors of the American evangelical online world emphasize the UN in their apocalyptic geopolitical texts on climate change and related issues.

As one of the most serious American evangelical students of geopolitical developments, Hal Lindsey (2015a) interprets climate change completely in geopolitical, apocalyptic, and conspiracist terms. And based on the established biblical knowledge of an apocalyptic one-world government, Lindsey mainly targets the UN as being complicit in establishing the global rule of the Antichrist. Yet, Herman (2001) states that Lindsey used to argue that the organizational structure of the UN obstructs the establishment of a global dictatorship and that, therefore, the UN is not of any eschatological significance. Rather, as Sturm and Albrecht (2021b) explain, Lindsey used to emphasize the EU (or earlier the European Economic Community) as a potential successor of the Roman Empire as well as Russia (or the USSR) in his construction of geopolitical *Feindbilder*. Moreover, an important post-Cold War transposition of Lindsey's prophetic geographies refocused to the Middle East, specifically Iran and Iraq.

Supporting that he perceives the UN usually as benign, Lindsey (2015a) claims on his TV and web show that the UN "has no teeth" but that the UN's leaders "would love to change that" with the UN's Climate Conference of 2015. Lindsey (2015a) believes that increased global governance is not a side effect or necessity to mitigate climate change, but the overarching goal of the deception of climate change, as he claims that the "ultimate goal of all that hype that is already surrounding this conference [COP21]" is "fear and control."

International climate change protection policies are of high relevance for Lindsey as they have the power to shift Lindsey's "prophetic geographies" from the EU or the Middle East toward the United Nations (Sturm, 2010: 142). Thereby, and rather than simply reconfirming existing prophetic geographies, climate change modifies the apocalyptic imagination of Hal Lindsey. As addressed in the conclusion of Chapter 5, Lindsey (2022) prophesied that after Covid, the global evil will focus again on climate change to subdue the world, showing that Lindsey still perceives climate change as prophetically relevant. Nonetheless, it must be noted that Lindsey adjusted his apocalyptic geopolitical imaginations to fit a contemporary trend within American evangelical dispensationalism, which is to perceive the UN as the NWO or the Antichrist system (Sutton, 2017; Barkun, 2010).

However, it is not just single conferences or agreements like that made at COP21 in 2015 which draw the attention of conspiracist evangelical prophecy writers. Climate change and connected discourses are frequently mentioned in the context of larger alleged anti-Christian superconspiracies, such as the UN's 2015 Sustainable Development Goals (SDG) or the Agenda 21. Within evangelical conspiracist environments, the SDG as well as the Agenda 21/30 are step-by-step guidelines that lay out how the Antichrists' global dictatorship is going to be established. For example, RaptureReady's Geri Ungurean (2016) writes concerning the SDG:

> As I read the agenda for the 2015 Sustainable Development Goals of the United Nations, I believe that the Lord opened my eyes and allowed me to

see with clarity that this meeting and its agenda was to prepare the way for the coming Antichrist.

Ungurean's article addresses all of the UN's documents more broadly and how it promotes global control, the reduction of the world's population, and economic policies, which are interpreted as "Marxism in its purest form" (Ungurean, 2016). The end goal of such Marxism is to create a "New World Order" that will subdue individual nation-states. Again, the historically established socialist/communist *Feindbild* of the American apocalyptic right (Herman, 2001; Werly, 1977; Hofstadter, 1964) is applied to the UN and sustainability/environmentalism, illustrating that decades-old (geo)political ideologies affect the perception of contemporary global non-binding action plans. Ungurean further argues that the 2015 UN Sustainable Development Goals are "clearly Satanic" and the "promise of the Antichrist system which will deceive the world." She concludes by stating that climate change policies are the UN's "all time favorite" propaganda, which is "without a doubt, the vehicle to the globalization of the world's economies."

EndTimeHeadlines frequently reposts articles from American alt-right websites, which similarly argue that Agenda 21/30, or the SDGs, are plans to impose a global socialist dictatorship and that global climate protection policies are an important element of these plans. Even though these articles do not imply any connection to the Antichrist or biblical End-Times, they are granted a religious meaning by virtue of being embedded in a religious digital space. Therefore, those rather non-religious texts contribute to reinforcing the religious anti-globalism bias of these digital apocalyptic environments by reinforcing their culturally compartmentalized beliefs by imbricating them with secular outsiders.

The UN's Agenda 2030 and climate change are central parts of the Covid-19 Great Reset conspiracy theory as well. Brentner (2020) writes on *RaptureForums* that "climate change and COVID-19 lie behind the push for the Great Reset. We see evidence of this in the United Nations' Agenda 2030, which lists goals for a one-world government that it hopes to put in place by 2030." What is remarkable here is that Brentner's text uses apocalyptic date setting, specifically 2030, as the UN's ambition to establish a one-world government. In a similar manner, the evangelical blog SignsOfEndTimes (no date) focuses on the year 2030 and claims that the UN intentionally picked the date 2030 to gain control of the world because the UN knows that Jesus will return about 2,000 years after he was crucified in AD 30–31. Referring to the UN's Agenda 2030 and Jesus' Second Coming, the authors of the blog write (2020):

> Now we are NOT setting any dates here, and neither should anyone say "Oh I have 10 years to get ready, plenty of time!", because as the apostle

Paul said, God will "finish the work, and cut it short in righteousness" (Romans 9:28). We are merely showing you that the second coming is "AT THE DOOR" (Matthew 24:33), and why there is such a push by the UN to gain control by 2030.

The practice of date setting is frowned upon among American evangelicals after numerous predictions of specific dates for the End-Times events failed and is a practice that has long been replaced by a sense of imminence brought on by geopolitical factors (Sturm, 2010). While the reference to the year 2030 does not resemble the numerological date-setting techniques of earlier prophecy writers, it somewhat provides an apocalyptic landmark for some American apocalypticists. It is a prophetically relevant date, conveniently imposed by outside forces. Therefore, the UN's Agenda 2030 is not only used to reconfirm the anti-globalism bias of American apocalypticism, but the supposed promise of the Antichrist's one-world government by the year 2030 also functions as an indicator of the imminent End-Times and thereby becomes embedded into dispensational eschatology.

Section II: American Exceptionalism— Resisting the Geographies of Control

Apocalyptic anti-globalist discourses within American evangelicalism are often accompanied by religious beliefs in an "elect" or special role of the United States. We refer to this as religiously motivated *American exceptionalism* (Barkun, 2010; Dittmer, 2010), a geopolitical imaginary which is reproduced and reinforced by ACC-skeptic thought. We argue that resistance to global climate protection policies not only reconfirms pre-existing anti-globalist sentiments but also, conversely, apocalyptic anti-globalist climate change discourses confirm beliefs in the distinctiveness and superiority of the United States. Drawing from imaginations of the United States as God's chosen nation (Gorski, 2019; Guth, 2012), several premillennial authors argue that only the United States can resist the anti-Christian climate change deception. Therefore, American evangelical ACC-skepticism constitutes a nationalist act of opposition toward what Barkun (2010: 126) calls the "geographies of control" of the apocalyptic NOW, which are based on the spatial imaginations of the United States as a divine space of resistance.

Dalby (2010: 99) writes that "American exceptionalism runs through American political culture and is evoked in numerous tropes in geopolitical discourse," including apocalyptic geopolitical discourse. American exceptionalist beliefs in the uniqueness and, often, superiority of the United

States in comparison to other nations are an essential part of the American evangelical identity and the American civil religion (Dittmer, 2010). In religious milieus, American exceptionalism is motivated by the popular religious imagination of the United States as a sacred place—a belief which is derived from evangelical millennialism. For example, during the colonization of America, postmillennialists thought that they were building God's kingdom on Earth and ushering in the Christian millennium through the establishment of a unique and exceptional Christian nation on sacred, American soil provided by God (Gribben, 2011; Marsden, 1980). In relation to the environment, Rozario (2001: 75) writes that:

> Puritans claimed to be especially gladdened by the fact that disasters destroyed property—reminding communities of the transience of worldly goods, freeing them form the distractions of material possessions, and refocusing their thoughts on the one thing that really mattered: salvation. Disasters, then, were blessings because they assisted colonists along the path to God's kingdom, inspiring moral and spiritual reformation, and promising them a final transcendence of space and time.

These rather postmillennial Puritans sought to make Heaven on Earth and sought to ensure an outcome of community and religious cohesion rather than individual salvation. Despite their individualism and fatalism, premillennialists also have an eschatological justification for American exceptionalism. Dittmer (2010) writes, for instance, that many premillennialist evangelicals interpret the United States as one of the Young Lions who fight against the evil powers Gog and Magog as described in Ezek. 38:13.[6] Many apocalyptic evangelicals see the United States as the Manichean counterpart of apocalyptic evil (Dalby, 2010; Dittmer, 2010). In the context of contemporary geopolitics and contemporary apocalyptic anti-globalism, it is believed that the United States must constitute a holdout to satanic globalist endeavors (Barkun, 2010).

The premillennial ideal of the United States as a space of resistance to satanic globalism became clear during the presidency of Donald Trump. His America First politics and his alleged rejection of globalism/globalization and criticism of intergovernmental treaties and institutions seemed to confirm the divinely designated uniqueness of the United States. Through isolationist policies and critiques of NATO, the UN, the World Bank, and the WHO, Trump satisfied the exceptionalist and isolationist beliefs of one of his most important electorates: white apocalyptic evangelicals. Consequently, when Trump announced in June 2017 that the United States was going to withdraw from the Paris Climate Agreement, apocalyptic evangelicals endorsed this announcement—the United States was once again chosen. In contrast, young and liberal evangelicals criticized the withdrawal and Trump's climate skeptic attitudes more broadly (Ricker, 2020; Stover, 2019).

American evangelical apocalyptic authors quickly provided prophetic interpretations of Trump's announcement regarding the Paris Climate Agreement and argued that the withdrawal not only protects American freedoms but also constitutes a prophetically relevant resistance to anti-Christian globalism. For example, in a YouTube video on the announced American withdrawal from the Paris Climate Agreement, Andy Woods and Jim McGowan claim that due to Trump's decision, "Satan's agenda for this world just suffered a massive setback" (Woods and McGowan, 2017). With all the bravado of American exceptionalism, Woods describes the United States as the "only real holdout" and stands in contrast to the EU's position on climate change. In relation to the EU's support of the Paris Climate Agreement, Woods explains that the "mind-set of Europe has acquiesced into a socialist, globalist . . . one world agenda described in Revelation 13." Here, Woods applies America's exceptionalist "holdout" status not only to climate change but also to the metric system and the Euro currency. According to the Texan pastors, it is part of America's DNA to resist Europe and globalism in virtually all aspects, as they argue that the American Founders wanted the United States to not participate in any global agreements—and by withdrawing from the Paris Agreement, Trump supposedly continues to comply with the will of the American Founders and the American Constitution.

In that context, Woods and McGowan (2017) mention George Washington, who, according to the Texan pastors, claimed that America should not go into permanent alliances with foreign nations since Washington has "expressed distrust of entangling foreign alliances." Conspiracist End-Times thought that prophesied a foreign invasion that destroys all American freedoms was used as a powerful resource by advocates of American independence (O'Leary, 1998; Boyer, 1992). Uscinski and Parent (2014: 3) state that the Founders of the United States assumed a British conspiracy against American settlers and further state that this "shaky conspiracy theory" contributed to justify the independence. Woods and McGowan are able to utilize the historic anti-globalism of the Founders to justify their premillennial support for Trump's decision to withdraw from the Paris Climate Agreement, as the first and only country to make such an "exceptional" announcement.

Moreover, the reference to the American Founders and the claim that Trump's decision is a setback for Satan constitute a "fusion of secular belief in American exceptionalism and religious narrative" (Dittmer, 2010: 80). RaptureReady's Daymond Duck (2018) argues in a similar manner that Trump has the power to "single-handedly destroying" the plans of the NWO after he "pulled the U.S. out of the Paris Climate Change Accords." Duck further argues that Russia, under its leader Vladimir Putin, fights the NWO together with the United States/Trump.

The portrayal of Russia as belonging to the "good" side in the fight against the anti-Christian NWO contradicts established premillennial geopolitical attitudes of the Cold War Era. It was presumed, even "gospel," that the USSR was Gog and Magog (Sutton, 2017; Gribben, 2011). Such a revisioning of Russia illustrates the flexibility of apocalyptic thought and its ability to adjust to political preferences and contemporary world politics (Howard, 2011). O'Donnell (2021) and Durbin (2020) observe that evangelical apocalypticists ascribe eschatological significance to Trump due to his political and anti-globalist attitudes, and as such, climate change must be understood as only one subject where Trump fights the globalist evil other. This somewhat challenges traditional accounts of evangelical prophecy, which usually teaches that the biblical End-Times are inevitable and cannot be stopped by humans like Trump. But since Trump is sometimes perceived as "God's Man in the White House" (Beverley, 2020), his anti-globalism can be constructed as being a predestined part of End-Times prophecy. The QAnon movement, which is endorsed by many apocalypticists of the American right, also constructs Trump as the chosen one who is able to combat and destroy the globalist Deep State (Bond and Neville-Shepard, 2021).

In that context, Gorski (2019) observes that the cult of personality surrounding Trump changed evangelicals' attitudes toward political leaders. Evangelicals usually criticize the divinization of political leaders as they only follow God and Jesus as saviors. But after "American exceptionalists have generally been immune to . . . political idolatry" (2019: 349), Trump became accepted by some conservative and right-wing evangelicals as something like a "god-king"—a political leader accorded with "divine intention" while acting in the interest of conservative and right-wing evangelicals (Durbin, 2020: 116). Trump's alleged ability to fight the Antichrist through ACC-skepticism shows that evangelicals perceive Trump's election as a "divine challenge to a demoniac status quo" (O'Donnell, 2020a: 713). Although focusing on Trump, the evangelical cult of personality and populism still reinforces American nationalism and exceptionalism embedded in American ACC-skeptic apocalyptic conspiracism: only the United States, "God's chosen nation," can have a leader powerful enough to make the Antichrist suffer a setback toward global power.

However, there are also evangelical adherents of conspiracism who do not believe that Trump's actions can somehow influence the approaching rise of the Antichrist and the imminent biblical End-Times. RaptureReady's administrator, Todd Strandberg (2017), called Trump's announcement in a post on RaptureForums a "huge victory for freedom loving Americans" as they "dodged the climate change bullet." Still, Strandberg claims that the United

States will not "have a respite from other similar entanglements" since prophecy predicts that "Satan will soon have control of all nations." In a similar manner, Hal Lindsey (2017) called Trump's decision only a "brief reprieve" because he believes that the Paris Climate Agreement is "a significant step toward global political control for European elites. That means, it is a step toward Antichrist." These dissonances within American evangelical digital spaces regarding Trump's decision to pull out of the Paris Climate Agreement illustrate that within the highly politicized discursive field of evangelical apocalypticism, some prophecy writers, like Woods and McGowan or Duck, are more flexible regarding the imminency of the End while others, like Strandberg and Lindsey, hold on to the timely End of Times regardless of possible prophetically relevant political "setbacks."

Trump and the withdrawal from the Paris Climate Agreement resonate with the religious exceptionalist belief of America as the new Jerusalem, where "God is apparently on America's side" (Dalby, 2010: 111) in the fight against global evil. Since the end of the Cold War that popularized the NWO belief (Barkun, 2010) and the Rio Earth Summit in 1992, American apocalypticists of the religious anti-globalist right propagated that climate change and global governance are tools of the Antichrist. The announcement of leaving the Paris Agreement in 2017 confirms the religious exceptionalist belief that America is a sacred space of resistance and a divine nation which fights the Antichrist.

Nonetheless, the active role of Trump in fighting the anti-Christian NWO somewhat contradicts the deterministic idea of evangelical apocalypticism. Premillennialism usually holds that humanity has no power to influence the ongoing apocalyptic fight between good and evil, which will lead to the biblical End-Times. Barkun (2010: 126) addresses this "millenarian tension between determinism and choice": it is assumed that "the rise of antichrist is a necessity of the end-time plan," which cannot be prevented, whereas at the same time, "the rallying cry to arm against the forces of Antichrist in order to protect American sovereignty paradoxically, is made by the same clergy who lay out the invariant millennial timetable." It is therefore important to recognize that "dispensationalists are not the fatalists they propound to be" and still "change history" as they claim that God guides their actions within his predetermined plan (Sturm, 2010: 133).

Such a change of history is created by the "mobilization and action against the forces of evil" through the endorsement of anti-globalist conspiracy beliefs, denouncement of environmentalism, and rejection of all political and epistemic institutions that support climate protection policies and environmentalism (Barkun, 2010: 126). In that context, the discursive construction of Christian America as "good" and God's chosen nation in contrast to the rest of the world as an apocalyptic "evil" must be understood as a performance of nationalism which is expressed through religious discourse: "religion *as* nationalism" (Sturm, 2017: 299). American conspiracists see themselves not only threatened by global climate protection policies and a satanic one-world government itself, but at the same time they fear the decay of the United States, the end of American sovereignty, and the loss of individual freedom secured by the American Constitution. Since religious NWO discourses on climate change claim that globalist conspirators strive to "create a world state in which individual nations and boundaries have disappeared," the resistance to climate change protection turns into a nationalist duty for adherents of End-Times conspiracism (Barkun, 2010: 126). It is believed that through political anti-environmentalism, support of conservative politics, anti-globalism, and anthropogenic climate change denial, the American nation can be protected. Apocalyptic conspiracist ACC-skepticism then constitutes a religious and, at the same time, nationalist practice which is intended to resist the loss of individual freedoms and national sovereignty.

Section III: The Evil Other within the United States and the Apocalyptic Demonization of the Domestic

While the American exceptionalist thought might suggest a simple bipolar Manichean geopolitical imagination prevalent in American digital evangelicalism (United States vs. the Antichrist's NWO), the primary sources on climate change we engage with illustrate that conspiracist End-Times discourses not only see evil others outside of the United States. Rather, certain authors discursively construct demonic enemies within the boundaries of the United States and within its sovereign control due, in part, to their position on climate change.

For conspiracist evangelicals, environmentalism and the belief in anthropogenic climate change indicate "globalist" tendencies and, thereby, an alignment with the Antichrist. Such constructions of evil are geopolitically relevant because they are fueled by apocalyptic anti-globalism and a form of religious nationalism which propagates an "us-vs-them" ideology in which even American supporters of globalism (them) are automatically perceived to be an enemy of Christian America (us). This construction of the enemy can be comprehended through the concept of demonization or demonology, which describes "a system of knowledge that arises to codify, classify and comprehend perceived threat to the essence and existence of an 'us'" (O'Donnell, 2020a: 700). Since conspiracism acts as an "early warning system for group security" (Uscinski and Parent, 2014: 17), it is employed as a powerful discursive tool in the construction of an apocalyptic, demonic threat.

Eventually, anti-globalist End-Times thought can "erect and reinforce barriers" between groups (Davis, 2018: 302) and make distinctions between the sacred and the profane. In the case of this research, this means constructed barriers between "anti-globalist" Americans and "globalist" Americans. What is important here is that the evangelical apocalyptic understanding of national identity does not depend on origin or citizenship but on geopolitical attitudes, which are indicated by the belief and disbelief in climate change. Consequently, we argue that for evangelical apocalyptic conspiracists, attitudes on climate change are an important factor in what we call, inspired by O'Donnell's (2020a, 2020b, 2019) research on apocalyptic evangelicals, the "apocalyptic demonisation of the domestic" which contributes to the American cultural divide, or as O'Donnell (2019) put it, the "End of One America."

In that context, the political left or democratic American politicians are frequently presented as globalist demons and outside threats to Christian America due to an assumed coalition with the Antichrist's NWO. For example, before Trump's presidency, Obama was demonized on RaptureReady as a "Marxist/Socialist New World Order puppet of the Elite" (Ungurean, 2015) due to his support of climate protection policies and other alleged globalist activities. In an EndTimeMinistries podcast, the host Dave Robbins (2019b) argues that former presidential candidate Bernie Sanders' Green New Deal "is really their efforts to implement the United Nation's socialistic to govern the world." In his two podcasts on climate change published in 2019, Robbins (2019a, 2019b) further targets Democrats Liz Warren, Alexandria Ocasio-Cortez, and Al Gore as well as center-left publications like *The New York Times*, *USA TODAY*, *The Washington Post*, and a *CNN* town hall event from 2019 in its entirety as globalist demons. Robbins argues that these politicians and media institutions are complicit in the promotion of the climate change hoax, which will bring the Antichrist into power. Virtually any democratic politician or media institution who advocates climate protection can be demonized as a globalist minion of the Antichrist.

Additionally, spatial entities are demonized as advancers of globalism. Greenfield (2017) on RaptureForums addresses California's declaration that it would continue to implement the Paris Climate Agreement in opposition to Trump's government. Greenfield interprets this announcement as a betrayal of the United States and contrasts globalist tendencies in relation to California, which, following an intertextual relationship to other texts on the same website, would indicate an alignment with the Antichrist. Here, Greenfield explains that individual "states cannot and are not allowed to unilaterally choose to "uphold" a treaty [Paris Climate Agreement] rejected by the President." The president's authority weighs heavily in the case of global climate change policies as Trump sought to preserve American independence, isolation, and sovereignty from global treaties. However, "the problem isn't limited to the Climate Alliance," as Greenfield accuses the Democratic Californian governor of "openly allying with China against the President of the United States" and further observes a connection between Democrats and Russia.

Contrasting a premillennial tradition of seeing the American federal government, as well as several American presidents, as controlled by demonic powers (O'Donnell, 2019; Sutton, 2012; Dittmer, 2010), Greenfield argues in favor of President Trump due to his divine popularity among evangelicals. This illustrates the transformability of premillennialism, in which the imaginations of a satanic enemy are adjusted according to contemporary political preferences. As soon as some characteristics of premillennialist demonic evil are fulfilled (cooperation with communist governments like China and Russia, environmentalism, and support for global treaties), "the process of making evil spaces of the Other" (Sturm, 2010: 136) in Greenfield's geographical imagination shifts to the Californian territory. A similar demonization of the state of California is found in a 2020 article on climate change, posted also on RaptureForums (2020), in which the author argues in a conspiracist manner that California intends to indoctrinate schoolchildren with climate protection and environmentalism. Even though the original article is authored by the rather non-religious alt-right website frontpagemag.com, it is embedded into an apocalyptic conspiracist digital environment and thereby reinforces an evangelical apocalyptic demonization of California.

It does not matter if it is a person or a state, evangelical apocalyptic anti-globalists can potentially perceive anyone and anything as a demonic outsider, a "them," an evil other, and a threat to their group security, if this person or spatial entity is believed to advance a globalist agenda. Whitehead and Perry (2020: 89) write in their assessment of Christian nationalism that "ambassadors of Christian nationalism are fond of boundaries—both physical and symbolic" and argue that Christian nationalists use a variety of factors to categorize their world into an "us" and a "them," for instance race, atheism, religiosity, conceptions of gender subordination, authorial control, and related topics. What is missing in their assessment, and what we argue is very important in light of the examples discussed in this section, are geopolitical attitudes. In that context, O'Donnell (2020a, 2019) states that apocalyptic Christians of the American right construct globalists as something demonic, regardless of whether these supposed globalists are of American or foreign origin. The *Feindbild* of an outside threat is projected onto Americans that threaten the idealized American way of life. American globalists function as a domestic surrogate for foreign enemies of Christian America, and the advocacy for climate protection and environmentalism helps apocalyptic conspiracists to identify domestic globalists.

Conclusion: Apocalyptic Conspiracism as Doing Geopolitics

This chapter explored anti-globalist geopolitical attitudes embedded in American evangelicals' apocalyptic conspiracist discourse on climate change. The dominant global view on climate change, as represented by the 195 signatories to the Paris Climate Agreement, is that climate change and global warming are significantly accelerated by human activity and issues, which require international cooperation and global governance to mitigate and adapt to the adverse effects of climate change. End-Times conspiracists reject that dominant view on climate change and utilize the internet to distribute geopolitical counter-elite discourses. Based on premillennial prophecy, it is claimed that a climate change deception will lead to a one-world government, and consequently, to the biblical End-Times. Nonetheless, similar narratives which postulate that the Antichrist strives for global control through worldwide conspiracies have existed for decades, or even centuries, within the milieu of American evangelical premillennialism. This longevity leads, for instance, Hermann (2001), Barkun (2013, 2010) or Sutton (2017) to observe a historical genealogy of anti-globalism bias within American evangelical premillennialism—and anthropogenic climate change denial became another part of this genealogy. Although there are significant intertextual gaps between the geopolitics of climate change and the End-Time verses of the Bible, conspiracist prophecy writers use the Bible to legitimize their conspiracist interpretations of climate change to reinforce their anti-globalist regime of truth and existing *Feindbilder*.

Conversely, climate change can be used as evidence of uniqueness or exceptionalism of the United States as God's chosen nation. After the United States, under Donald Trump, announced the withdrawal from the Paris Climate Agreement, evangelicals saw the exceptionalist nature of the United States reconfirmed. Trump's decision concerning the Paris Climate Agreement was not universally accepted and was later overturned by his successor, Joe Biden—a Democrat and supporter of global climate protection policies and, therefore, a potential demon for apocalyptic conspiracist evangelicals. O'Donnell (2020a, 2019) argues that Christian nationalists demonize Americans who supposedly align themselves with globalists, and this chapter's research on climate change confirms this demonization of the domestic based on anti-globalist geopolitical imaginations.

Moreover, this chapter's research was intended to stress the geopolitical dimension of apocalyptic conspiracist discourse as an alternative, counter-elite understanding of global power relations. Such discourses must be discussed as a cultural, social, and political practice (Müller, 2008; Ó Tuathail and Dalby,

1998), which "draws an alternative reality of global politics that challenge Liberal, Realist, and Marxist geopolitical approaches and that explain global politics in terms of scheming, power-obsessed conspirators" (Albrecht and Sturm, 2021: 7). It is important here to acknowledge such conspiracist geopolitical discourse as "a knowledge-producing discourse" (Jones, 2012: 45) and as "abstract forms of knowledge" (Müller, 2008: 329) since they claim to know how the world *will* look like while uncovering secret plots.

What we have shown is that apocalypticism and conspiracism, with their prophetic characteristics, are suitable vessels for geopolitical discourses since "geopolitics is often associated with futurology and is used to predict future global schisms and shape of power, whether concerned with energy competition, ideological clashes or conflicts over civilisations" (Dittmer and Dodds, 2008: 438). Here, the internet, as an "important medium for the anti-globalization movement" (Dodds, 2007: 120), is used to distribute their prophetic predictions and interpretations of world politics. Apocalyptic geopolitical counter-knowledge is produced in evangelical apocalyptic digital spaces, as "geopolitics has different sites of production" (Ó Tuathail and Dalby, 1998: 5). Apocalyptic or conspiracist explanatory models of geopolitics constitute an easily accessible and sustainable simplification of the impalpable reality of global politics and the "messy complexity of the world" (Jones, 2012: 53). Dittmer and Dodds (2008) explain that a function of geopolitical discourse is to divide global space into simplistic categories. Today, American apocalyptic conspiracism, with its Manichean characteristics, manages to divide space and render "the global" as evil and Christian America as the "good." The imagination of "the global" as demonic (O'Donnell, 2020a, 2019) is the most recent form of "the dominant Other in the American geopolitical imaginary" (Dalby, 2008: 418). Still, these anti-globalist geopolitical discourses have the ability to "shape how people understand the world around them and, thus, how they choose to act within it" (Jones, 2012: 45). American millennialism is a powerful political force in the United States (Agnew, 2006) and has the ability to produce "bizarre and dangerous geopolitical scenarios with real political effects" (Megoran, 2012: 146)—and when looking at Trump's presidency, his ties with apocalyptic evangelicals (however opportunistic) and secular alt-right conspiracists, QAnon Shamans, and the insurrection upon the Capitol building, it can be said that the real-life political effects of End-Times conspiracism have been significant . . . and "bizarre" (2012).

8

Conclusion

Apocalyptic Friends and the Truth of the End

In times when conspiracist ways of knowing become increasingly visible and powerful as they motivate parts of the population to denounce climate protection policies (Lewandowsky et al., 2018; Jolley and Douglas, 2014), to reject potentially life-saving vaccines (Röchert et al., 2021; Allington et al., 2021), or to storm governmental buildings (Lee et al., 2022; Bond and Neville-Shepard, 2021), this book explored the cultural histories and epistemic practices that yield politically radical, science-skeptic, and government-skeptic conspiracy beliefs. We engaged with a highly politicized way of *knowing* about global crises that is the result of established epistemic strategies and counter-knowledge of the religious and secular American anti-globalist right. It was a central concern of this book project to take the conspiracists and their counter-knowledge seriously, not to denounce or debunk their respective ways of knowing (following Harambam, 2021; Hagen, 2020), but to ask *how* and *why* certain groups in the United States, mainly politically conservative or right-wing apocalyptic evangelicals, generate and distribute truth-claims that hold that our lives are governed by evil global superpowers.

This said, and to attempt self-reflexivity, our own presuppositions and positions inevitably show through our analysis of the social construction of counter-truths in digital spaces. Just like the authors we analyzed, our texts reflect the epistemic strategies and authorities we trust, be it the IPCC or philosophers and academics like Michel Foucault or Michael Barkun, whose prior knowledge that we accept as legitimate caused us to interpret evangelical apocalyptic discourse in a certain way. Inspired by Barkun's

(2013) *improvisational millennialism* and Robertson's (2016; 2018) *millennial conspiracism* as well as academic discussions on the intersectionality of American millennialism and conspiracy beliefs (e.g., O'Donnell, 2021, 2020a; Wilson, 2017; Dittmer, 2010; O'Leary, 1998; Boyer, 1992; Hofstadter, 1964), we employed the concept of *apocalyptic conspiracism* to address conspiracy discourses that predict disaster, suffering, and the end of the contemporary world.

Addressing the growing popularity and power of conspiracy beliefs in contemporary societies, Barkun (2018: ix) writes that "the boundary between truth and falsity has become blurred for many people." Blurred boundaries can exist for those who possibly believe in a few populist counter-knowledge claims or endorse some conspiracy theories while still following mainstream beliefs on the majority of issues. From the perspective of apocalyptic conspiracists, there are usually clear boundaries between truths and lies. Their respective apocalyptic version of the truth rarely acknowledges any uncertainties because of their established "mechanisms and instances which enable one to distinguish true and false statements" (Foucault, 1980: 131). The conspiracist way of knowing explored in this book is based on a form of epistemic dualism, or epistemic Manicheanism, which generates clear boundaries between apocalyptic truths only the elect can access and lies and deceptions that the mainstream has fallen for (DiTommaso, 2020).

The End-Times discourses discussed above frequently follow a holistic worldview—the idea that everything is connected, that nothing occurs accidentally, and that evil forces control the entire human destiny through superconspiracies (Aupers and Harambam, 2018; Wilson, 2017; Barkun, 2013). Religious as well as secular apocalyptic conspiracists emphasize not only the individual events that happen but also the "interrelationships between the various elements of something" (Wood and Douglas, 2018: 94). For instance, Covid vaccinations, mask mandates, the papal encyclical letter *Laudato Si'*, democratic proposals for a Green New Deal, Al Gore's *An Inconvenient Truth*, and the Paris Climate Agreement are believed to be part of the same malicious geopolitical superconspiracy. All are amalgamated into sound-bite terms—Agenda 21, Agenda 30, New World Order, or the Great Reset—that at once stand in to mean everything and nothing as throwaway exemplars.

In this holistic worldview, individual truth-claims and attendant *Feindbilder* (e.g., Pope supports the NWO due to his environmentalism, Bill Gates cooperates with the NWO due to his involvement in the WHO) reinforce each other, constantly legitimizing the overall apocalyptic conspiracist worldview and creating a self-sustaining and non-falsifiable system of counter-knowledge (Barkun, 2013; Howard, 2006). Based on the geopolitical belief in an End-Times superconspiracy, adherents argue that individual crises like climate change prepare the world for the satanic one-world government, while at the same time, the trueness of the geopolitical superconspiracy seems to be confirmed by the existence of the individual conspiracist plots. There is a reciprocal relationship between the single conspiracy theories (at the micro-level) and the geopolitical conspiracism (at the macro-level). Each digital trace, like an article, a video, or podcast, ascribing recent events to the End-Times superconspiracy, contributes to sustaining the apocalyptic conspiracist "regime of truth" (Foucault, 1980) in an intertextual manner, as no piece of discourse emerges isolated from previous knowledge in the same cultural environment or digital space (Hodges, 2015; Paltridge, 2012). Geography matters!

Political Enemies, Apocalyptic Friends

All of the evangelical and secular apocalyptic conspiracist truths explored in this book have a specific functional objective: to challenge the political and epistemic institutions in power and to initiate political and societal change (Robertson, 2016; Jones, 2012; Levy, 2007). The political activism of the evangelical side of the apocalyptic conspiracist alliance seems at first glance paradoxical. Although premillennialism holds that humanity's downfall, accompanied by the rise of evil, is inevitable, evangelical apocalypticists still actively participate in the political discourse in the United States and try to shape public opinions (Sutton, 2012; Sturm, 2010; Dittmer, 2007). In conspiracist spaces—digital or physical—debates between different ways of knowing do not exist for the sake of it but rather have clear functional aims of political change, or at least the avoidance of change. In the case of climate change, this means that any sustainable policies like the Green New Deal or global climate protection policies like the Paris Climate Agreement need to be prevented. It is not the mainstream knowledge or the epistemic authorities themselves, but the social and political effects of certain knowledge which motivate conspiracists to start websites, publish articles, distribute podcasts, or storm governmental buildings. Here, Foucault (1980: 82) states:

> We are concerned with the insurrection of knowledges that are opposed primarily not to the contents, methods or concepts of a science, but to the effects of the centralising powers which are linked to the institutions and functioning of an organised scientific discourse within a society such as ours.

In other words, it is the political ramifications of certain knowledge-claims and the related accumulation of power by political institutions that drive contemporary End-Times conspiracists (Harambam and Aupers, 2015). This is the case with some American evangelicals who do not oppose science itself, as often wrongly assumed, but certain forms of science, and in particular those with concrete political and societal consequences, like climate science (Evans, 2011; Roberts, 2008).

Therefore, apocalyptic conspiracist discourses are politically relevant knowledge, a powerful alternative imagination of the truth. And these perceptions of truth and lies are connected to moral judgments about what is right and what is wrong. End-Times conspiracism is a practice which strives to establish an alternative truth as the dominant truth and to bring about societal and political change.

To conclude this book, we wish to return to an argument made in the introduction and highlight that apocalyptic thought in combination with certain political purposes are exclusive to the American right. Climate change and environmental degradation can create the impression that human civilization is on the brink of disaster, that Earth is becoming uninhabitable, and that the planet seems to be in a state of irreversibly slow decay. The end of the world can feel palpable, present, and imminent. As addressed in the introduction of this book, Greta Thunberg stated that "the world is on fire" at the 2020 World Economic Forum in Davos in her address titled, "Averting a Climate Apocalypse." Similarly, Al Gore compares climate change and natural disasters to the book of Revelation. Many environmentalists employ apocalyptic language or imaginaries to communicate climate science and to warn against the consequences of global warming and climate change (Skrimshire, 2013; Swyngedouw, 2013, 2010; Hulme, 2009). Moreover, the names of two of the most popular environmentalist movements in recent years, Extinction Rebellion and Fridays For Future, carry an apocalyptic tone as they suggest that mass extinction is imminent (and that it must be rebelled against) and that the entire future of younger generations is at stake. Without action, there will not be a future, only end of time. Therefore, secular environmentalists also draw from religious apocalypticism in their discursive framing of the climate- or eco-apocalypse, for instance, through imaginations of Earth or nature as a divine entity. Here, "narratives of climate change as the power of 'mother Nature' applying its logic against human schemes" (Skrimshire, 2014: 243) reflect beliefs in a supernatural punishment as a result of human misbehavior, or sin. Indeed, radical geographers like Swyngedouw (2010) claim that the use of apocalypse within secular climate discourses has been captured by capitalism to limit democratic input into policy change. Implicitly, the "parallel culture amid the interstices of the secular world" (Stephens and Giberson, 2011: 180) among climate apocalypse narratives suggests "this is an age of variegated and competing climate apocalypses" (Sturm and Lustig, 2022: 213). Of course, missing from much of this discussion are the more progressive and theoretically driven evangelical voices critical of apocalypse conceptually on the one hand, but equally attempting to refigure it within its redemptive political impetus. Keller (1999: 57–8) writes in relation to apocalyptic climate change visions:

> Apocalypse means unveiling, disclosure. So we had better tease its contemporary incarnations out of their bitter sense of closure, toward their own dis/closive potential. Counter/apocalypse allows us to strengthen the relation of the apocalyptic unconditional to the relativizing, relational conditionals of the larger biblical tradition . . . in the process perhaps emergency gives way to emergence, and uncertainty to adventure.

In contrast to the Bible, where "that day and that hour knoweth no man, no, not the angels which are in heaven, neither the Son, but the Father" (Mk 13:32)[1] climate science can provide a quantification of the always imminent End-Times (e.g., a rise in global temperature by a certain year). An IPCC (2018) report published in 2018 presented climate models which show that emissions need to be reduced drastically by 2030 to limit the rise of the global average temperature to 1.5°C. This report led several media outlets and environmental organizations like Extinction Rebellion to claim that there are only "12 years to save Earth" (Kreps, 2018). Such date-setting has a long history in apocalyptic movements. Likewise, *The Guardian* published a headline which stated, "We have 12 years to limit climate change catastrophe, warns UN" (Watts, 2018). But what is often neglected in the public communication of climate science is that the IPCC (2018) highlights the inaccuracies, interquartile ranges, and statistical limitations of its climate models[2] so that a statement like "we only have 12 years left" does not fully capture and represent the "post-normal" characteristics of global climate change knowledge. The complexities and uncertainties of climate science are undermined by apocalyptic framings and certain prophecies of impending disaster at a given time. In addition, scientific terms like "tipping points" or "thresholds" implicate an abrupt transformation for the worse, a point of no return, or an apocalypse (Hulme, 2009). Rode and Fischbeck (2021: 204) address such "forecasts of an apocalyptic event" and note "that neglecting uncertainty results in greater visibility, but shifts focus away from prognosticators with scientific backgrounds" (2021: 207).

Overall, apocalyptic rhetoric acts as a simplifier of climate science where "debates over 'sustainable' futures in the face of pending environmental catastrophe signal a range of populist maneuvers that infuse the post-political post-democratic condition" (Swyngedouw, 2010: 221). The popular and populist imaginings of an imminent climate apocalypse impose a global politics of consensus and political action. Nonetheless, oversimplifications of climate science as well as the connected apocalyptic rhetoric in the public communication of climate models, often denoted as hype and alarmism by critics, can subvert the credibility of climate science as a whole (Rode and Fischbeck, 2021). Some climate advocates even construct an apocalyptic good-versus-evil dualism in which someone is either considered an environmentalist who tries to avert the climate catastrophe or a climate sinner because, for many, "every action and political identity is given cosmic significance by playing a part within a dualistic framework" (Skrimshire, 2014: 243). To remind us of DiTommaso's (2020: 318) statement concerning apocalyptic dualisms: "One cannot 'agree to disagree' when eternal life (or equivalent) is at stake."

Some of the environmental movement, climate science, and activism embrace apocalypticism in several ways: the language, the doomsaying, the good/evil dualism, and other ideas that are borrowed from Christian apocalyptic concepts. Based on such characteristics, the premillennial authors from RaptureReady (no date) denote environmentalists as "our apocalyptic friends." This potentially contradictory statement suggests that premillennial evangelicals somewhat perceive environmentalists as apocalypticists, but as ones that do not possess the hidden knowledge and real truth about the future. This conclusion is captured succinctly in the full quote: "Despite the fact that there is zero evidence supporting their position, our apocalyptic friends continue to ramp up the lies."

Both apocalyptic American evangelicals and apocalyptic environmentalists believe they are in possession of the real truth about the future, and even though they both focus on climate change to predict/prophesize the end of times, their political goals could not be more different. Each group tries to establish its perception of the truth as the dominant one to accomplish their respective political goals as "we are subjected to the production of truth through power and we cannot exercise power except through the production of truth" (Foucault, 1980: 93). Environmentalist movements like Fridays For Future call for pervasive and global political actions to mitigate and adapt to the adverse effects of climate change, and thereby, to prevent the apocalypse and instead initiate a millennium of sustainability and climate justice. Yet, these pervasive global political responses constitute part of the apocalypse for premillennial evangelicals. Still, environmentalist anthropogenic climate change knowledge follows the epistemic authorities, while the premillennial conspiracist response counters such dominant knowledge (Levy, 2007). Certainly, the climate protection movement is not devoid of conspiracist knowledge-claims when, for example, some politicians or oil companies are accused of covering up the adverse effects of greenhouse gas emissions in an attempt to normalize fossil fuel use (Rowlands, 2000). But what counts as a conspiracy theory for some can be the truth and moral obligation for others, and what counts as the prosperous millennium for some is the apocalypse for others, and this judgment always works both ways.

Trump and Dominant Epistemic End-Times

Apocalyptic conspiracism is more than a temporary trend. Former president Donald Trump's 2024 re-election campaign has fomented further climate skepticism and conspiracies, pushing it further into the grip of the far-right and to his evangelical base. Trump's grip on the Republican Party means that any deviant voices from within the party who are not populist, far-right enough, or conspiracist enough, are marginalized. As such, he has largely unified his party through apocalyptic conspiracism, and as a result, such conspiracism is now part of the American mainstream via media, culture, religion, and politics. Without presenting our own prophecies and cautioning against alarmist apocalypticism, many polls at the time of writing in early 2024 indicate another Trump win, but of course, much can happen before the presidential election. In primary rallies and interviews, Trump has not shied away from declaring his intention to dismantle environmental legislation like clean energy tax credits and international climate agreements, as he did with the Paris Agreement. Waldman (2024) expects "a second Trump presidency to show less restraint" on dismantling US climate policy architecture. Recently, Trump has pandered to and influenced his evangelical denialist base with conspiracist language that referred to climate change as a "make-believe problem," "non-existent," and a "hoax." In a repositioning of apocalyptic threats from climate to an arms race, Trump announced in December 2023 that "the only global warming we should be worrying about . . . is nuclear warming."

Apocalyptic conspiracism is a powerful tool not just for political motivation but also for identity formation. Through feelings of both being elect to special knowledge to identify outside values and forces, but also social cohesion of victimization. On this, Perry (2024: np) notes that "One of the central reasons evangelicals have rallied to Trump again is because they are not only partisans but culture warriors who feel under attack." Regardless of whether Trump is in office or will be, through three presidential election races, he has managed to position himself and his electorate as the suppressed and victimized people who are in an earthly and cosmic fight with leftist evil. Through a counter-epistemic position of suppression under Biden, evangelicals can convincingly continue to sell this discourse. Here, the conspiracy theory of Trump's stolen 2020 election also confirmed that evil powers attempt to do everything to suppress God's chosen president.

Learning from Trump's presidency, we should take what Trump says seriously; he has shown he will act on his words. His prosperity gospel-driven and opportunistic relationship with evangelicals (given that they are more likely than any other group to deny climate change and support Trump) would almost certainly mean he will continue to chip away at environmental protection and climate action. While certainly climate change denialism is biblically driven, we also outline in this book that evangelical skepticism is also political. As Veldman (2019: 216) concludes, skepticism is "not simply what happened when evangelicals considered the issue in light of scripture, but in part the result of certain evangelical leaders and pundits having an interest in portraying skepticism as the scriptural view."

The 900-page document "Project 2025" by the conservative think tank *Heritage Foundation* also illustrates the subjectivity and plasticity of conspiracism. The document outlines methods that intend to reshape the US executive and judicial branches during Trump's desired next presidency in favor of a conservative and right-wing agenda. In short, the plan, created by groups closely linked to Trump's camp, describes ways of removing existing constitutional restraints to accommodate the implementation of Trump's goals in different levels of governance. According to Project 2025, a Trump government should place sympathizers in relevant political and public positions, like stacking the courts with climate change skeptics, to allow Trump to further roll back environmental legislation. Project 2025 itself, a plan that aims to install Trump's minions in all dimensions of public and political life, can be perceived as a conspiracy masterplan of right-wing populism.

The difference here between the apparent hidden evil leftist New World Order and the conservative Project 2025 is that this blueprint for a conservative, or rather right-wing and authoritarian, leadership is published in plain sight. Yet, Project 2025 uses the same methods that conspiracists ascribe to the NWO regime: give power to sympathizers of the preferred political agenda to affect public opinions and create a new political reality.

Apocalyptic conspiracism about anthropogenic climate change will motivate and be motivating politics in the United States in this election and beyond. Understanding how it functions and for whom will be crucial to understand if we intend to prevent such counter-epistemic discourse from becoming the dominant episteme.

Notes

Chapter 4

1. [Dan. 7:23-24 KJV] (23) Thus he said, The fourth beast shall be the fourth kingdom upon earth, which shall be diverse from all kingdoms, and shall devour the whole earth, and shall tread it down, and break it in pieces. (24) And the ten horns out of this kingdom are ten kings that shall arise: and another shall rise after them; and he shall be diverse from the first, and he shall subdue three kings.
2. [Mt 24:4-5] "And Jesus answered and said unto them, Take heed that no man deceive you." (5) "For many shall come in my name, saying, I am Christ; and shall deceive many."
3. [2 Jn 1:7] "For many deceivers are entered into the world, who confess not that Jesus Christ is come in the flesh. This is a deceiver and an antichrist."
4. [Rev. 19:20] "And the beast was taken, and with him the false prophet that wrought miracles before him, with which he deceived them that had received the mark of the beast, and them that worshipped his image. These both were cast alive into a lake of fire burning with brimstone."
5. [2 Thess. 2:9] "Even him, whose coming is after the working of Satan with all power and signs and lying wonders."

Chapter 5

1. Like all of Lindsey's publications, *The Hal Lindsey Report* should be seen as part of a larger premillennial knowledge network in which different authors develop ideas, ghost-write texts, and publish material under Lindsey's name. A possible origin of Lindsey's prophetic interest in the politics of climate change could be his former ghostwriter, Jack Kinsella. Kinsella was the principal writer for *The Hal Lindsey Report* in the 1990s and early 2000s. Before his death in 2013, Kinsella repeatedly ascribed high prophetic significance to global climate politics. On Kinsella's own blog, Omega Letter (omegaletter.com), Kinsella and his respective writers and associates published dozens of apocalyptic conspiracist articles on global warming and environmental politics between 2004 and 2018. After Kinsella's passing, *Omega Letter* continued to publish articles on climate change, including reposts of articles Kinsella authored before his death. Lindsey's prophetic knowledge-claims concerning global

warming and environmental politics resemble many of Kinsella's arguments made on *Omega Letter*, so that it could have been Kinsella's legacy that influenced Lindsey's perception of climate change.

2 [Gen. 1:28] "And God blessed them, and God said unto them, Be fruitful, and multiply, and replenish the earth, and subdue it: and have dominion over the fish of the sea, and over the fowl of the air, and over every living thing that moveth upon the earth."

3 "Even if it turns out to be real and harmful, is of little ultimate importance, compared to how one is 'going to live [either in heaven or hell for] eternity.'" Beisner quoted in Zaleha and Szasz (2014: 216).

4 Beisner recommended books on amillennialism twice on the Cornwall Alliance's blog, while I could not identify any support of either premillennialism or postmillennialism (Beisner, 2015).

5 [Rom. 1:20-25] (20) For the invisible things of him from the creation of the world are clearly seen, being understood by the things that are made, even his eternal power and Godhead; so that they are without excuse: (21) Because that, when they knew God, they glorified him not as God, neither were thankful; but became vain in their imaginations, and their foolish heart was darkened. (22)Professing themselves to be wise, they became fools, (23 And changed the glory of the uncorruptible God into an image made like to corruptible man, and to birds, and fourfooted beasts, and creeping things. (24) Wherefore God also gave them up to uncleanness through the lusts of their own hearts, to dishonour their own bodies between themselves: (25) Who changed the truth of God into a lie, and worshipped and served the creature more than the Creator, who is blessed for ever. Amen.

Chapter 6

1 [Gen. 9:8-13] (8) And God spake unto Noah, and to his sons with him, saying, (9) And I, behold, I establish my covenant with you, and with your seed after you; (10) And with every living creature that is with you, of the fowl, of the cattle, and of every beast of the earth with you; from all that go out of the ark, to every beast of the earth. (11) And I will establish my covenant with you, neither shall all flesh be cut off any more by the waters of a flood; neither shall there any more be a flood to destroy the earth. (12) And God said, This is the token of the covenant which I make between me and you and every living creature that is with you, for perpetual generations: (13) I do set my bow in the cloud, and it shall be for a token of a covenant between me and the earth.

Chapter 7

1 [Dan. 7:23] Thus he said, The fourth beast shall be the fourth kingdom upon earth, which shall be diverse from all kingdoms, and shall devour the whole

earth, and shall tread it down, and break it in pieces. [Dan. 7:24] And the ten horns out of this kingdom are ten kings that shall arise: and another shall rise after them; and he shall be diverse from the first, and he shall subdue three kings.
2. [Lk. 21:25] And there shall be signs in the sun, and in the moon, and in the stars; and upon the earth distress of nations, with perplexity; the sea and the waves roaring.
3. [Rev. 13:7] And it was given unto him to make war with the saints, and to overcome them: and power was given him over all kindreds, and tongues, and nations.
4. [Rev. 12:9] "And the great dragon was cast out, that old serpent, called the Devil, and Satan, which deceiveth the whole world: he was cast out into the earth, and his angels were cast out with him."
5. [Lk. 21:25] "And there shall be signs in the sun, and in the moon, and in the stars; and upon the earth distress of nations, with perplexity; the sea and the waves roaring"
6. [Ezek.38:13] "Sheba, and Dedan, and the merchants of Tarshish, with all the young lions thereof, shall say unto thee, Art thou come to take a spoil? hast thou gathered thy company to take a prey? to carry away silver and gold, to take away cattle and goods, to take a great spoil?"

Chapter 8

1. Although the Bible can be interpreted to argue against date-setting practices in Mk 13:32 (or also Mt. 25:13), several pre-millennial Christian prophecy writers used numerology or have attempted to set a date for the biblical End-Times and Jesus' Second Coming. Due to repeated failures in predicting the apocalypse and the biblical command to refrain from date-setting, the practice became increasingly unpopular in the twentieth century (Boyer, 1992).
2. In 2018, the IPCC (2018) described a modeled scenario in which the increase in global temperatures is limited to 1.5°C (compared to pre-industrial levels). This scenario requires that "global net anthropogenic CO_2 emissions decline by about 45% from 2010 levels by 2030 (40–60% interquartile range), reaching net zero around 2050 (2045–2055 interquartile range)" (IPCC, 2018: 12).

References

Agnew, J. (2006), "Religion and Geopolitics," *Geopolitics*, 11 (2): 183–91.
Agrawala, S. (1998), "Structural and Process History of the Intergovernmental Panel on Climate Change," *Climatic Change*, 39: 621–42.
Ahmed, W., Vidal-Alaball, J., Downing, J., and Seguí, F. L. (2020), "COVID-19 and the 5G Conspiracy Theory: Social Network Analysis of Twitter Data," *Journal of Medical Internet Research*, 22 (5), [online]. Available at: https://www.jmir.org/2020/5/e19458/
Albrecht, T. and Sturm, T. (2021), "Conspiratorial Geopolitics and COVID-19 Counter-Epistemic Knowledge Mutations in Germany," in T. Sturm, J. Mercille, T. Albrecht, J. Cole, K. Dodds, and A. Longhurst (eds.), Interventions in Critical Health Geopolitics: Borders, Rights, and Conspiracies in the COVID-19 Pandemic." *Political Geography*, 91: 102445.
Alder, C. and Schäublin, E. (2020), "US Evangelicals: From Prophecy to Policy," *CSS Policy Perspectives*, 8 (11): 1–4.
Allington, D., McAndrew, S., Moxham-Hall, V., and Duffy, B. (2021), "Coronavirus Conspiracy Suspicions, General Caccine Attitudes, Trust and Coronavirus Information Source as Predictors of Vaccine Hesitancy among UK Residents During the COVID-19 Pandemic," *Psychological Medicine*, [online]. Available at: https://doi.org/10.1017/S0033291721001434.
Amoore, L. (2020), *Cloud Ethics: Algorithms and the Attributes of Ourselves and Others*. Durham: Duke University Press.
Amos, C., Spears, N., and Pentina, I. (2016), "Rhetorical Analysis of Resistance to Environmentalism As Enactment of Morality Play between Social and Ecological Well-Being," *The Journal of Consumer Affairs*, 50 (1): 224–59.
Andersen, K., Shehata, A., and Andersson, D. (2021), "Alternative News Orientation and Trust in Mainstream Media: A Longitudinal Audience Perspective," *Digital Journalism*, 11 (1): 1–20.
Arbuckle, M. B. (2017), "The Interaction of Religion, Political Ideology, and Concern About Climate Change in the United States," *Society & Natural Resources*, 30 (2): 177–94.
Armaly, M. T., Buckley, D. T., and Enders, A. M. (2022), "Christian Nationalism and Political Violence: Victimhood, Racial Identity, Conspiracy, and Support for the Capitol Attacks," *Political Behavior*, 44: 937–60.
Asad, T. (1993), *Genealogies of Religion: Discipline and Reasons of Power in Christianity and Islam*, Baltimore: The Johns Hopkins University Press.
Asad, T. (2003), *Formations of the Secular: Christianity, Islam, Modernism*. Stanford: Stanford University Press.
Ash, J., Kitchin, R., and Leszczynski, A. (2018), "Digital Turn, Digital Geographies?" *Progress in Human Geography*, 42 (1): 25–43.

Asprem, E. and Dyrendal, A. (2018), "Close Companions? Esotericism and Conspiracy Theories," in A. Dyrendal, D. G. Robertson, and E. Asprem (eds.), *Handbook of Conspiracy Theory and Contemporary Religion*, 207–33, Leiden: Brill.

Aupers, S. (2012), "Trust No One: Modernization, Paranoia and Conspiracy Culture," *European Journal of Communication*, 27 (1): 22–34.

Aupers, S. and Harambam, J. (2018), "Rational Enchantments: Conspiracy Theory between Secular Scepticism and Spiritual Salvation," in A. Dyrendal, D. G. Robertson, and E. Asprem (eds.), *Handbook of Conspiracy Theory and Contemporary Religion*, 48–69, Leiden: Brill.

Bakshy, E., Messing, S., and Adamic, L. A. (2015), "Exposure to Ideologically Diverse News and Opinion on Facebook," *Science*, 348 (6239): 1130–2.

Barker, D. C. and Bearce, D. H. (2012), "End-Times Theology, the Shadow of the Future, and Public Resistance to Addressing Global Climate Change," *Political Research Quarterly*, 66 (2): 267–79.

Barkun, M. (1986 [1974]), *Disaster and the Millennium*, Syracuse University Press edn, originally published. New Haven: Yale University Press.

Barkun, M. (1996), *Religion and the Racist Right: The Origins of the Christian Identity Movement*, Revised edn, Chapel Hill: The University of North Carolina Press.

Barkun, M. (2010), "The 'New World Order' and American Exceptionalism," in J. Dittmer and T. Sturm (eds.), *Mapping the End Times: American Evangelical Geopolitics and Apocalyptic Visions*, 119–32, Surrey: Ashgate Publishing Limited.

Barkun, M. (2013 [2003]), *A Culture of Conspiracy: Apocalyptic Visions in Contemporary America*, 2nd edn, Berkeley: University of California Press.

Barkun, M. (2016), "Conspiracy Theories as Stigmatized Knowledge," *Diogenes*, 62 (3–4): 114–20.

Barkun, M. (2018), "Foreword," in A. Dyrendal, D. G. Robertson, and E. Asprem (eds.), *Handbook of Conspiracy Theory and Contemporary Religion*, ix–x, Leiden: Brill.

Barr, S. (2011), "Climate Forums: Virtual Discourses on Climate Change and the Sustainable Lifestyle," *Area*, 43 (1): 14–22.

Bastos, M., Mercea, D., and Baronchelli, A. (2018), "The Geographic Embedding of Online Echo Chambers: Evidence from the Brexit Campaign," *PloS one*, 13 (11), [online]. Available at: https://doi.org/10.1371/journal.pone.0206841

Baum, M. A. and Groeling, T. (2008), "New Media and the Polarization of American Political Discourse," *Political Communication*, 25 (4): 345–65.

Bauman, W. (2014), *Religion and Ecology: Developing a Planetary Ethic*, New York: Columbia University Press.

Bebbington, D. W. (1989), *Evangelicalism in Modern Britain: A History from the 1730s to the 1980s*, London: Unwin Hyman.

Beck, S. and Mahony, M. (2018), "The IPCC and the New Map of Science and Politics," *WIREs Clim Change*, 9 (6): 1–16.

Beesley, D. (2011), "YouTube and Apocalyptic Rhetoric: Broadcasting Yourself to the Ends of the World," in R. G. Howard (ed.), *Network Apocalypse: Visions of the End in an Age of Internet Media*, 44–73, Sheffield: Sheffield Phoenix Press.

Bekkering, D. J. (2011), "From 'Televangelist' to 'Intervangelist': The Emergence of the Streaming Video Preacher," *The Journal of Religion and Popular Culture*, 23 (2): 101–17.

Bergmann, S. (2009), "Invoking the Spirit amid Dangerous Environmental Change," in S. Bergmann (ed.), *Religion, Space and the Environment*, 159–74, Minneapolis, MN: Lutheran University Press.

Bergquist, P. and Warshaw, C. (2019), "Does Global Warming Increase Public Concern About Climate Change?" *The Journal of Politics*, 81 (2): 686–91.

Berlet, C. (2004), "Christian Identity: The Apocalyptic Style, Political Religion, Palingenesis and Neo-Fascism," *Totalitarian Movements and Political Religions*, 5 (3): 469–506.

Berry, D. T. (2017), *Blood and Faith: Christianity in American White Nationalism*, Syracuse, NY: Syracuse University Press.

Beverley, J. (2020), *God's Man in the White House: Donald Trump in Modern Christian Prophecy*, Burlington: Castle Quay Books.

Bevir, M. (1999), "Foucault, Power, and Institutions," *Political Studies*, 47 (2): 345–59.

Bjerg, O. and Presskorn-Thygesen, T. (2017), "Conspiracy Theory: Truth Claim or Language Game?" *Theory, Culture & Society*, 34 (1): 137–59.

Bjurström, A. and Polk, M. (2011), "Physical and Economic Bias in Climate Change Research: A Scientometric Study of IPCC Third Assessment Report," *Climatic Change*, 108 (1–2): 1–22.

Bloom, M. and Rollings, R. (2022), "Losing My Religion: Evangelicalism and the Gospel of Q," *Journal of Religion and Violence*, 10 (1): 1–15.

Bolin, B. (2001), "Politics and the IPCC," *Science*, 296 (5571): 1235.

Bond, B. E. and Neville-Shepard, R. (2021), "The Rise of Presidential Eschatology: Conspiracy Theories, Religion, and the January 6th Insurrection." *American Behavioral Scientist*, [online]. Available at: https://doi.org/10.1177/000276422110465

BonJour, L. (2002), "Internalism and Externalism," in P. K. Moser (ed.), *The Oxford Handbook of Epistemology*, 234–63, Oxford: Oxford University Press.

Bourdieu, P. (1998), *Practical Reason*, Stanford: Stanford University Press.

Boyer, P. S. (1992), *When Time Shall Be No More- Prophecy Belief in Modern American Culture*, Cambridge, MA: The Belknap Press of Harvard University Press.

Brace, C. and Geoghegan, H. (2010), Human Geographies of Climate Change: Landscape, Temporality, and Lay Knowledges," *Progress in Human Geography*, 35 (3): 284–302.

Bratich, J. Z. (2008), *Conspiracy Panics: Political Rationality and Popular Culture*, Albany: Suny Press.

Bricker, B. J. (2013), "Climategate: A Case Study in the Intersection of Facticity and Conspiracy Theory," *Communication Studies*, 64 (2): 218–39.

Bolin, B. (2001), "Politics and the IPCC," *Science*, 296 (5571): 1235.

Bray, D. and Martinez, G. (2015), "Climate-Change Lore and Its Implications for Climate Science: Post-Science Deliberations?" *Futures*, 66 (2015): 54–69.

Brint, S. and Abrutyn, S. (2010), "Who's Right About the Right? Comparing Competing Explanations of the Link Between White Evangelicals and Conservative Politics in the United States," *Journal for the Scientific Study of Religion*, 49 (2): 328–50.

Bruns, A., Harrington, S., and Hurcombe, E. (2020), "Corona? 5G? or both?: The Dynamics of COVID-19/5G Conspiracy Theories on Facebook," *Media International Australia*, 177 (1): 12–29.

Burke, M. (2000), *A Social History of Knowledge*, Malden, MA: Blackwell Publishing Company.

Butler, J. (1990) *Gender Trouble: Feminism and the Subversion of Identity*. London: Routledge.
Campbell, H. A. (2017), "Surveying Theoretical Approaches within Digital Religion Studies," *New Media & Society*, 19 (1): 15–24.
Carr, W., Patterson, M., Yung, L., and Spencer, D. (2012), "The Faithful Skeptics: Evangelical Religious Beliefs and Perceptions of Climate Change," *Journal for the Study of Religion, Nature & Culture*, 6 (3): 276–99.
Carrion, M. L. (2018), "'You Need to do Your Research': Vaccines, Contestable Science, and Maternal Epistemology," *Public Understanding of Science*, 27 (3): 310–24.
Carroll, B. E. (2012), "Worlds in Space: American Religious Pluralism in Geographic Perspective," *Journal of the American Academy of Religion*, 80 (2): 304–64.
Chaudoin, S., Smith, D. T., and Urpelainen, J. (2014), "American Evangelicals and Domestic Versus International Climate Policy," *The Review of International Organizations*, 9 (4): 441–69.
Climate Feedback. (2019), "Non-peer-reviewed Manuscript Falsely Claims Natural Cloud Changes can Explain Global Warming," *Climate Feedback Reviews*, Available at: https://climatefeedback.org/claimreview/non-peer-reviewed-manuscript-falsely-claims-natural-cloud-changes-can-explain-global-warming/ (Last Accessed: June 25, 2022).
Coaston, J. (2018), "YouTube, Facebook, and Apple's ban on Alex Jones, Explained," *Vox*, August 6, Available at: https://www.vox.com/2018/8/6/17655658/alex-jones-facebook-youtube-conspiracy-theories (Last Accessed: June 25, 2022)
Cobb Jr., J. B. (2004), "Protestant Theology and Deep Ecology," in R. S. Gottlieb (ed.), *This Sacred Earth: Religion, Nature, Environment*, 248–61, New York and London: Routledge.
Cohn, N. (1970 [1957]), *The Pursuit of the Millennium*, Revised and Expanded edn, Oxford: Oxford University Press.
Collins, J. J. (2020), "Apocalypticism as a Worldview in Ancient Judaism and Christianity," in C. McAllister (ed.), *The Cambridge Companion to Apocalyptic Literature*, 19–36, Cambridge: Cambridge University Press.
Concise Oxford English Dictionary. (2011), "Apocalypse," Twelfth edn, Oxford: Oxford University Press.
Connolly, W. E. (2008), *Capitalism and Christianity, American Style*, Durham: Duke University Press.
Cook, J. (2017), "Understanding and Countering Climate Science Denial," *Journal and Proceedings of the Royal Society of New South Wales*, 150 (465/466): 207–19.
Cook, J. (2020), "Deconstructing Climate Science Denial," in D. C. Holmes and L. M. Richardson (eds.), *Research Handbook on Communicating Climate Change*, 62–78, Cheltenham: Edward Elgar Publishing.
Coverley, D. M. (2017), *Global Warming or God's Warming: A Prophetic Explanation for the Strange and Unusual Events in the Skies, on the Land, in the Waters, and with the Weather*, Bloomington: WestBow Press.
Craig, E. (1990), *Knowledge and the State of Nature*, Oxford: Clarendon Press.
Crossley, J. (2021), "The Apocalypse and Political Discourse in an Age of COVID," *Journal for the Study of the New Testament*, 44 (1): 93–111.

Curry-Roper, J. M. (1990), "Contemporary Christian Eschatologies and their Relation to Environmental Stewardship," *The Professional Geographer,* 42 (2): 157–69.

Dahlgren, P. (2005), "The Internet, Public Spheres, and Political Communication: Dispersion and Deliberation," *Political Communication,* 22 (2): 147–62.

Dalby, S. (2008), "Imperialism, Domination, Culture: The Continued Relevance of Critical Geopolitics," *Geopolitics,* 13 (3): 413–36.

Dalby, S. (2010), "Critical Geopolitics and Security," in J. P. Burgess (ed.), *The Routledge Handbook of New Security Studies,* 50–8, London: Routledge.

Dalby, S. (2013), "The Geopolitics of Climate Change," *Political Geography,* 37: 38–47.

Dallek, M. (2023), *Birchers: How the John Birch Society Radicalized the American Right,* New York: Basic Books.

Danielsen, S. (2013), "Fracturing Over Creation Care? Shifting Environmental Beliefs Among Evangelicals, 1984-2010," *Journal For The Scientific Study of Religion,* 52 (1): 198–215.

Davis, M. (2005), *The Monster at Our Door: The Global Threat of Avian Flu,* New York: New Press.

Dayton, D. W. and Johnston, R. K. (1991), "Introduction," in D. W. Dayton and R. K. Johnston (eds.), *The Variety of American Evangelicalism,* 1–4, Knoxville: The University of Tennessee Press.

Demeritt, D. (2001), "The Construction of Global Warming and the Politics of Science," *Annals of the Association of American Geographers,* 91 (2): 307–37.

De Pryck, K. and Gemenne, F. (2017), "The Denier-in-Chief: Climate Change, Science and the Election of Donald J. Trump," *Law and Critique,* 28 (2): 119–26.

de Zeeuw, D., Hagen, S., Peeters, S., and Jokubauskaite, E. (2020), "Tracing Normiefication: A Cross-platform Analysis of the QAnon Conspiracy Theory," *First Monday,* 25 (11) [online]. Available at: https://doi.org/10.5210/fm.v25i11.10643

DiMaggio, A. R. (2022), "Conspiracy Theories and the Manufacture of Dissent: QAnon, the 'Big Lie', Covid-19, and the Rise of Rightwing Propaganda," *Critical Sociology,* [online]. Available at: https://doi.org/10.1177/08969205211073669

DiTommaso, L. (2020), "Apocalypticism in the Contemporary World," in McAllister, C. (ed.), *The Cambridge Companion to Apocalyptic Literature,* 316–42, Cambridge: Cambridge University Press.

Dittmer, J. (2007), "Of Gog and Magog: The Geopolitical Visions of Jack Chick and Premillennial Dispensationalism," *ACME: An International E-Journal for Critical Geographies,* 6 (2): 278–303.

Dittmer, J. (2010), "Obama, Son of Perdition?: Narrative Rationality and the Role of the 44th President of the United States in the End-of-days," in J. Dittmer and T. Sturm (ed.), *Mapping the End Times: American Evangelical Geopolitics and Apocalyptic Visions,* 73–95, Surrey: Ashgate Publishing Limited.

Dittmer, J. and Dodds, K. (2008), "Popular Geopolitics Past and Future: Fandom, Identities and Audiences," *Geopolitics,* 13 (3): 437–57.

Dittmer, J. and Spears, Z. (2009), "Apocalypse, Now? The Geopolitics of 'Left Behind'," *GeoJournal,* 74 (3): 183–9.

Dittmer, J. and Sturm, T. (2010), *Mapping the End Times: American Evangelical Geopolitics and Apocalyptic Visions.* Surrey: Ashgate Publishing Limited.

Djupe, P. A. and Gwiasda, G. W. (2010), "Evangelizing the Environment: Decision Process Effects in Political Persuasion," *Journal for the Scientific Study of Religion*, 49 (1): 73–86.
Dochuk, D. (2011), *From Bible Belt to Sunbelt: Plain-Folk Religion, Grassroots Politics, and the Rise of Evangelical Conservatism*, New York: Norton.
Dochuk, D. (2019), *Anointed with Oil: How Christianity and Crude Made Modern America,* New York: Basic Books.
Dodds, K. (2007), *Geopolitics: A Very Short Introduction*, New York: Oxford University Press.
Douglas, C. (2021), "Revenge Is a Genre Best Served Old: Apocalypse in Christian Right Literature and Politics," *Religions*, 13 (1) [online]. Available at: https://doi.org/10.3390/rel13010021
Douglas, K. M. and Sutton, R. M. (2015), "Climate Change: Why the Conspiracy Theories are Dangerous," *Bulletin of the Atomic Scientists,* 71 (2): 98–106.
Duplaga, M. (2020), "The Determinants of Conspiracy Beliefs Related to the COVID-19 Pandemic in a Nationally Representative Sample of Internet Users," *International Journal of Environmental Research and Public Health*, 17 (21) [online]. Available at: https://doi.org/10.3390/ijerph17217818
Durbin, S. (2018), *Righteous Gentiles: Religion, Identity, and Myth in John Hagee's Christians United for Israel*, Leiden: Brill.
Durbin, S. (2020), "From King Cyrus to Queen Esther: Christian Zionists' Discursive Construction of Donald Trump as God's Instrument," *Critical Research on Religion,* 8 (2): 115–37.
Dutton, W. H., Blank, G., and Groselj, D. (2014), "Cultures on the Internet," *InterMEDIA,* 42 (4/5): 55–7.
Ecklund, E. H., Scheitle, C. P., Peifer, J., and Bolger, D. (2017), "Examining Links between Religion, Evolution Views, and Climate Change Skepticism," *Environment and Behavior*, 49 (9): 985–1006.
Evans, J. H. (2011), "Epistemological and Moral Conflict Between Religion and Science," *Journal for the Scientific Study of Religion,* 50 (4): 707–27.
Evans, J. H. and Feng, J. (2013), "Conservative Protestantism and Skepticism of Scientists Studying Climate Change," *Climatic Change*, 121 (4): 595–608.
Faddoul, M., Chaslot, G., and Farid, H. (2020), "A Longitudinal Analysis of YouTube's Promotion of Conspiracy Videos," *arXiv preprint* [online]. Available at: https://arxiv.org/abs/2003.03318
Farmer, G. T. and Cook, J. (2013), "Understanding Climate Change Denial," in G. T. Farmer and J. Cook (ed.), *Climate Change Science: A Modern Synthesis*, 445–66, Dordrecht: Springer.
Fitzgerald, T. (2011), *Religion and Politics in International Relations: The Modern Myth*, London and New York: Continuum.
Flaherty, E., Sturm, T., and Farries, E. (2022), "The Conspiracy of Covid-19 and 5G: Spatial Analysis Fallacies in the Age of Data Democratization," *Social Science & Medicine*, 293 (2022) [online]. Available at: https://doi.org/10.1016/j.socscimed.2021.114546
Foley, K. M., Ferrigno, J. G, Swithinbank, C, Williams Jr.,R. S., and Orndorff, A. L. (2013) "Coastal-change and glaciological map of the Amery Ice Shelf area, Antarctica: 1961–2004." USGS. Available at: https://pubs.usgs.gov/imap/2600/Q/ (Last Accessed: June 14, 2024).
Foucault, M. (1970 [1966]), *The Order of Things*, New York: Vintage Books.

Foucault, M. (1980), *Power/Knowledge: Selected Interviews and Other Writings 1972–1977*, Brighton: The Harvester Press.

Foucault, M. (1990), *The History of Sexuality, Volume 1: An Introduction* [Translated from the French by Robert Hurley], London: Penguin Books Ltd.

Frazier-Crawford Boerl, C. W. and Perkins, C. (2011), "The Political Pluralisation of American Evangelicals: How Old Media Built a Movement, and Why the Internet is Poised to Change it," *International Journal for the Study of the Christian Church*, 11 (1): 66–78.

Fricker, M. (1998), "Rational Authority and Social Power: Towards a Truly Social Epistemology," *Proceedings of the Aristotelian Society*, 98: 159–77.

Fricker, M. (2007), *Epistemic Injustice: The Power and Ethics of Knowing*, Oxford: Oxford University Press.

Fricker, M. (2010), "Scepticism and the Genealogy of Knowledge: Situating Epistemology in Time," in A. Haddock, A. Millar, A. Pritchard (eds.), *Social Epistemology*, 51–68, Oxford: Oxford University Press.

Fricker, M. (2012), "Group Testimony? The Making of a Collective Good Informant," *Philosophy and Phenomenological Research*, 84 (2): 249–76.

Friedrichs, J. (2011), "Peak Energy and Climate Change: The Double Bind of Post-Normal Science," *Futures*, 43 (2011): 469–77.

Frykholm, A. J. (2004), *Rapture Culture: Left Behind in Evangelical America*, New York: Oxford University Press.

Fuller, S. (2002), *Social Epistemology*, 2nd edn, Bloomington, IN.: Indiana University Press.

Gallagher, E. V. (2012), "Reading Prophetically: Millennialism, Prophecy and Tradition," in J. Searle and K. G. C. Newport (eds.), *The Future of Millennial Studies*, 148–67, Sheffield: Sheffield Phoenix Press.

Galvin, R. and Healy, N. (2020), "The Green New Deal in the United States: What it is and How to Pay for it," *Energy Research & Social Science*, 67 (2020) [online]. Available at: https://doi.org/10.1016/j.erss.2020.101529

Gerten, D. and Bergmann, S. (2012), "Facing the Human Faces of Climate Change," in D. Gerten and S. Bergmann (eds.), *Religion in Environmental and Climate Change- Suffering, Values, Lifestyles*, 3–15, London: Continuum International Publishing Group.

Geschke, D., Lorenz, J., and Holtz, P. (2019), "The Triple-filter Bubble: Using Agent-based Modelling to Test a Meta-theoretical Framework for the Emergence of Filter Bubbles and Echo Chambers," *British Journal of Social Psychology*, 58 (1): 129–49.

Ginzel, A. and Stoll, U. (2020), "Mit Judenstern gegen Corona-Maßnahmen," *ZDF Frontal 21*, May 19. Available at: https://www.zdf.de/politik/frontal/rechte-kapern-hygiene-demos-100.html (Last Accessed: June 21, 2022).

Goldman, A. I. (1999), *Knowledge in a Social world*, Oxford: Clarendon.

Goldman, A. I. (2002), *Pathways to Knowledge- Private and Public*, Oxford: Oxford University Press.

Goldman, A. I. (2010), "Why Social Epistemology is Real Epistemology," in A. Haddock, A. Miller, and D. Pritchard (eds.), *Social Epistemology*, 1–28, Oxford: Oxford University Press.

Gorski, P. (2019), "Why Evangelicals Voted for Trump: A Critical Cultural Sociology," in J. Mast and J. Alexander (eds.), *Politics of Meaning/Meaning of Politics. Cultural Sociology*, 165–83, Cham: Palgrave Macmillan.

Gottlieb, R. S. (2004), "Saving the World: Religion and Politics in the Environmental Movement," in R. S. Gottlieb (ed.), *This Sacred Earth: Religion, Nature, Environment*, 568–95, New York and London: Routledge.
Gottlieb, R. S. (2006), *A Green Faith*, Oxford and New York: Oxford University Press.
Gough, N. (2021), "The Christian Right's War on Reality: Where do/should American Science Teachers Stand?," *Cultural Studies of Science Education*, 16 (2): 419–28.
Gounaridis, D. and Newwll, J. P. (2024), "The Social Anatomy of Climate Change Denial in the United States," *Scientific Reports* 14: 2097.
Graham, S. (1998), "The End of Geography or the Explosion of Place? Conceptualizing Space, Place and Information Technology," *Progress in Human Geography*, 22 (2): 165–85.
Graham, M. (2014), "Internet Geographies: Data Shadows and Digital Divisions of Labour," in M. Graham and W. H. Dutton (eds.), *Society and the Internet: How Networks of Information and Communication are Changing our Lives*, 99–116, Oxford: Oxford University Press.
Graham, M. (2020), "Thomas Berry and the Reshaping of Catholic Environmentalism," *Worldviews: Global Religions, Culture, and Ecology*, 24 (2): 156–83.
Gregory, D., Meusburger, P., and Suarsana, L. (2015), "Power, Knowledge, and Space: A Geographical Introduction," in P. Meusburger, D. Gregory, and L. Suarsana (eds.), *Geographies of Knowledge and Power. Knowledge and Space*, Vol. 7, 1–18, Dordrecht: Springer.
Gribben, C. (2004), "Rapture Fictions and the Changing Evangelical Condition," *Literature and Theology*, 18 (1): 77–94.
Gribben, C. (2009) *Writing the Rapture: Prophecy Fiction in Evangelical America*. New York: Oxford University Press.
Gribben, C. (2011), *Evangelical Millennialism in the Trans-Atlantic World, 1500–2000*, Basingstoke: Palgrave Macmillan.
Gribben, C. (2024), *J.N. Darby and the Roots of Dispensationalism*, New York: Oxford University Press.
Guth, J. L. (2012), "The Religious Roots of Foreign Policy Exceptionalism," *The Review of Faith & International Affairs*, 10 (2): 77–85.
Hagemeister, M. (2018), "The Third Rome Against the Third Temple: Apocalypticism and Conspiracism in Post-Soviet Russia," in A. Dyrendal, D. G. Robertson, and E. Asprem (eds.), *Handbook of Conspiracy Theory and Contemporary Religion*, 423–42, Leiden: Brill.
Hagen, K. (2020), "Should Academics Debunk Conspiracy Theories?" *Social Epistemology*, 34 (5): 423–39.
Hahn, U., Harris, A. J., and Corner, A. (2016), "Public Reception of Climate Science: Coherence, Reliability, and Independence," *Topics in Cognitive Science*, 8 (1): 180–95.
Hamilton, M. L. C. (2024), Trumpism, Climate and COVID: Social Bases of the New Science Rejection," *PLoS ONE*, 19 (1): e0293059.
Hampel, M. (2016), "Have Climate Sceptics Taken the Bait? What the Deconstruction of Instrumental Climate Records Can Tell Us About the Politics of Climate Change," *Area,* 48 (2): 244–8.
Harambam, J. (2021), "Against Modernist Illusions: Why We Need More Democratic and Constructivist Alternatives to Debunking Conspiracy Theories," *Journal for Cultural Research*, 25 (1): 104–22.

Harambam, J. and Aupers, S. (2015), "Contesting Epistemic Authority: Conspiracy Theories on the Boundaries of Science," *Public Understanding of Science*, 24 (4): 466–80.

Harrington, J. (2009), "Evangelicalism, Environmental Activism, and Climate Change in the United States," *Journal of Religion and Society*, 11 (2009) [online]. Available at: https://dspace2.creighton.edu/xmlui/handle/10504/64433

Harvey, J. A., Van Den Berg, D., Ellers, J., Kampen, R., Crowther, T. W., Roessingh, P., Verheggen, B., Nuijten, R. J., Post, E., Lewandowsky, S., and Stirling, I. (2018), "Internet Blogs, Polar Bears, and Climate-change Denial by Proxy," *BioScience*, 68 (4): 281–7.

Herman, D. (2000), "The New Roman Empire: European Envisionings and American Premillennialists," *Journal of American Studies*, 34 (1): 23–40.

Herman, D. (2001), "Globalism's 'Siren Song': The United Nations and International Law in Christian Right Thought and Prophecy," *The Sociological Review*, 49 (1): 56–77.

Hodges, A. (2015), "Intertextuality in Discourse," in D. Tannen, E. Heidi, H. E. Hamilton, and D. Schiffrin (eds.), *The Handbook of Discourse Analysis*, 42–60, West Sussex: John Wiley & Sons.

Hofstadter, R. (1964), "The Paranoid Style in American Politics," *Harper's Magazine*, 77–86.

Holder, R. W. and Josephson, P. B. (2020), "Donald Trump, White Evangelicals, and 2020: A Challenge for American Pluralism," *Society*, 57 (5): 540–6.

Hopewell, D. (1998), *Christian Fundamentalism: A Journey into the Heart of Darkness*, Parsippany, NJ: American Atheist Press.

Howard, R. G. (2006), "Sustainability and Narrative Plasticity in Online Apocalyptic Discourse after September 11, 2001," *Journal of Media and Religion*, 5 (1): 25–47.

Howard, R. G. (2010), "Enacting a Virtual 'ekklesia': Online Christian Fundamentalism as Vernacular Religion," *New Media & Society*, 12 (5): 729–44.

Howard, R. G. (2011), *Digital Jesus: The Making of a New Christian Fundamentalist Community on the Internet*, New York: New York University Press.

Hsu, C. P. (2015), "Effects of Social Capital on Online Knowledge Sharing: Positive and Negative Perspectives," *Online Information Review*, 39 (4): 466–84.

Hughes, H. (2015), "The IPCC and the New Map of Science and Politics," *Global Environmental Politics*, 15 (4): 85–104.

Hulme, M. (2009), *Why We Disagree About Climate Change*, Cambridge: Cambridge University Press.

Hulme, M. (2011), "Reducing the Future to Climate: A Story of Climate Determinism and Reductionism," *Osiris*, 26 (1): 245–66.

Hulme, M. (2015a), "Varieties of Religious Engagement with Climate Change," in M. Tucker, W. Jenkins, and J. Grim (eds.), *Routledge Handbook of Religion and Ecology*, 239–48, London: Routledge.

Hulme, M. (2015b), "Finding the Message of the Pope's Encyclical," *Environment: Science and Policy for Sustainable Development*, 57 (6): 16–19.

Hulme, M. (2019), "Climate Emergency Politics is Dangerous," *Issues in Science and Technology*, 36 (1), 23–5.

Hulme, M. and Mahony, M. (2010), "Climate Change: What Do We Know About the IPCC?," *Progress in Physical Geography*, 34 (5): 705–18.

Hummel, D. G. (2016), "Revivalist Nationalism since World War II: From 'Wake up, America!' to 'Make America Great Again'," *Religions*, 7 (11) [online]. Available at: https://doi.org/10.3390/rel7110128

Hummel, D. G. (2018), *Covenant Brothers: Evangelicals, Jews, and U.S.-Israeli Relations*, Philadelphia: University of Pennsylvania Press.

Hummel, D. G. (2020), "American Evangelicals and the Apocalypse," in C. McAlister (ed.), *The Cambridge Companion to Apocalyptic Literature*, 288–316, Cambridge: Cambridge University Press.

Iheka, C. (2017), "Pope Francis' Integral Ecology and Environmentalism for the Poor," *Environmental Ethics*, 39 (3): 243–59.

Imhoff, R. and Lamberty, P. (2020), "A Bioweapon or a Hoax? Link Between Distinct Conspiracy Beliefs About the Coronavirus Disease (COVID-19) Outbreak and Pandemic Behavior," *Social Psychological and Personality Science*, 11 (8): 1110–18.

Inhofe, J. (2012), *The Greatest Hoax: How the Global Warming Conspiracy Threatens Your Future*, Chicago: WND Books.

IPCC. (2013), "Summary for Policymakers," in T. F. Stocker, D. Qin, G.-K. Plattner, M. Tignor, S. K. Allen, J. Boschung, A. Nauels, Y. Xia, V. Bex, and P. M. Midgley (eds.), *Climate Change 2013: The Physical Science Basis. Contribution of Working Group I to the Fifth Assessment Report of the Intergovernmental Panel on Climate Change*, 3–29, Cambridge: Cambridge University Press.

IPCC. (2018), "Summary for Policymakers," in V. Masson-Delmotte, P. Zhai, H.-O. Pörtner, D. Roberts, J. Skea, P. R. Shukla, A. Pirani, W. Moufouma-Okia, C. Péan, R. Pidcock, S. Connors, J. B. R. Matthews, Y. Chen, X. Zhou, M. I. Gomis, E. Lonnoy, T. Maycock, M. Tignor, and T. Waterfield (eds.), *Global Warming of 1.5°C. An IPCC Special Report on the Impacts of Global Warming of 1.5°C above Pre-Industrial Levels and Related Global Greenhouse Gas Emission Pathways, in the Context of Strengthening the Global Response to the Threat of Climate Change, Sustainable Development, and Efforts to Eradicate Poverty*. Cambridge: Cambridge University Press.

IPCC. (2022), "Summary for Policymakers," [H.-O. Pörtner, D. C. Roberts, E. S. Poloczanska, K. Mintenbeck, M. Tignor, A. Alegría, M. Craig, S. Langsdorf, S. Löschke, V. Möller, A. Okem (eds.)], in H.-O. Pörtner, D. C. Roberts, M. Tignor, E. S. Poloczanska, K. Mintenbeck, A. Alegría, M. Craig, S. Langsdorf, S. Löschke, V. Möller, A. Okem, B. Rama (eds.), *Climate Change 2022: Impacts, Adaptation, and Vulnerability. Contribution of Working Group II to the Sixth Assessment Report of the Intergovernmental Panel on Climate Change*. Cambridge: Cambridge University Press. In Press.

Jasanoff, S. (1997), "NGOs and the Environment: From Knowledge to Action," *Third World Quarterly*, 18 (3): 579–94.

Jasanoff, S. (2010), "A New Climate for Society," *Theory, Culture & Society*, 27 (2–3): 233–53.

Jasanoff, S. (2013), "A World of Experts: Science and Global Environmental Constitutionalism," *Environmental Affairs*, 40 (439): 439-452.

Jasanoff, S. and Simmet, H. R. (2017), "No Funeral Bells: Public Reason in a "Post-truth' Age," *Social Studies of Science,* 45 (5): 751–70.

Jebeile, J. (2020), "Values and Objectivity in the Intergovernmental Panel on Climate Change," *Social Epistemology*, 34 (5): 453–68.

Jenkins, W. (2009), "After Lynn White: Religious Ethics and Environmental Problems," *Journal of Religious Ethics,* 37 (2): 283–309.

Jenkins, W., Berry, E., and Kreider, L. B. (2018), "Religion and Climate Change," *Annual Review of Environment and Resources*, 43: 85–108.

John, S. (2017), "From Social Values to P-Values: The Social Epistemology of the Intergovernmental Panel on Climate Change," *Journal of Applied Philosophy*, 34 (2): 157–71.

Johnson-Schlee, S. (2019), "Playing Cards Against the State: Precarious Lives, Conspiracy Theories, and the Production of 'Irrational' Subjects," *Geoforum*, 101 (2019): 174–81.

Jolley, D. and Douglas, K. M. (2014), "The Social Consequences of Conspiracism: Exposure to Conspiracy Theories Decreases Intentions to Engage in Politics and to Reduce One's Carbon Footprint," *British Journal of Psychology*, 105 (1): 35–56.

Jolley, D. and Paterson, J. L. (2020), "Pylons Ablaze: Examining the Role of 5G COVID-19 Conspiracy Beliefs and Support for Violence," *British Journal of Social Psychology*, 59 (3): 628–40.

Jones, D. A. (2004), "Why Americans Don't Trust the Media: A Preliminary Analysis," *Harvard International Journal of Press/Politics*, 9 (2): 60–75.

Jones, L. (2012), "The Commonplace Geopolitics of Conspiracy." *Geography Compass*, 6 (1): 44–59.

Kaplan, A. (2018), *Our American Israel: The Story of an Entangled Alliance*, Cambridge: Harvard University Press.

Kappas, M. (2009), *Klimatologie: Klimaforschung im 21. Jahrhundert – Herausforderung für Natur- und Sozialwissenschaften*, Heidelberg: Spektrum Akademischer Verlag.

Katzav, J. and Parker, W. S. (2018), "Issues in the Theoretical Foundations of Climate Science," *Studies in History and Philosophy of Modern Physics*, August, 63: 141–9.

Kaufhold, K., Valenzuela, S., and De Zúñiga, H. G. (2010), "Citizen Journalism and Democracy: How User-generated News Use Relates to Political Knowledge and Participation," *Journalism & Mass Communication Quarterly*, 87 (3–4): 515–29.

Kauppinen, J. and Malmi, P. (2019), "No Experimental Evidence for the Significant Anthropogenic Climate Change," *arxiv.org*. Available at: https://doi.org/10.48550/arXiv.1907.00165

Kearns, L. (2012), "Religious Climate Activism in the United States," in D. Gerten and S. Bergmann (eds.), *Religion in Environmental and Climate Change - Suffering, Values, Lifestyles*, 132–51, London: Continuum International Publishing Group.

Kearns, L. (2014), "Green Evangelicals," in B. Steensland and P. Goff (eds.), *The New Evangelical Social Engagement*, 157–99, Oxford and New York: Oxford University Press.

Keeley, B. (2007), "God as the Ultimate Conspiracy Theory," *Episteme*, 4 (2): 135–49.

Keller, C. (1999), "The Heat is On: Apocalyptic Rhetoric and Climate Change," *Ecotheology*, 7 (1): 40–58.

Kidd, T. S. (2016), *Polls Show Evangelicals Support Trump. But the Term 'Evangelical' Has Become Meaningless*, July 22. Available at: https://www.washingtonpost.com/news/acts-of-faith/wp/2016/07/22/polls-show-evangelicals-support-trump-but-the-term-evangelical-has-become-meaningless/ (Last Accessed: June 25, 2022).

Knowles, S. (2013), "Rapture Ready and the World Wide Web: Religious Authority on the Internet," *Journal of Media and Religion*, 12 (3): 128–43.

Kovach, S. D. (2012), "God's Word in Human Words: An Evangelical Appropriation of Critical Biblical Scholarship," *Journal of the Evangelical Theological Society*, 55 (1): 208.

Kowalewski, J., eds. (2023), *The Environmental Apocalypse: Interdisciplinary Reflections on the Climate Crisis*, London: Routledge.

Kreps, D. (2018), "Federal Climate Change Report Warns of Bleak Economic, Societal Impact," *Rolling Stone*, November 24. Available at: https://www.rollingstone.com/politics/politics-news/federal-climate-change-report-759793/ (Last Accessed: June 25, 2022).

Kuhn, T. (2012 [1962]), *The Structures of Scientific Revolutions*, 4th edn, Chicago, IL: Chicago University Press.

Lahn, B. and Sundqvist, G. (2017), "Science as a 'Fixed Point'? Quantification and Boundary Objects in International Climate Politics," *Environmental Science & Policy*, 67: 8–15.

Landes, R. (2008), "Millennialism," in J. R. Lews (ed.), *The Oxford Handbook of New Religious Movements, Oxford Handbooks Online*. Oxford University Press. Available at: https://www.oxfordhandbooks.com/view/10.1093/oxfordhb/9780195369649.001.0001/oxfordhb-9780195369649-e-15?q=Christ (Last Accessed: December 28, 2021).

Latour, B. (2004) "Why has Critique Run out of Steam? From Matters of Fact to Matters of Concern," *Critical Inquiry*, 30 (2): 225–48.

Latour, B. and Woolgar, S. (1979), *Laboratory Life: The Social Construction of Scientific Facts*, Beverly Hills, CA: Sage.

Lee, C. S., Merizalde, J., Colautti, J. D., An, J., and Kwak, H. (2022), "Storm the Capitol: Linking Offline Political Speech and Online Twitter Extra-Representational Participation on QAnon and the January 6 Insurrection," *Frontiers in Sociology,* 7, 876070 [online]. Available at: https://doi.org/10.3389/fsoc.2022.876070

Lee, F. L. (2015) "Internet Alternative Media use and Oppositional Knowledge," *International Journal of Public Opinion Research*, 27 (3): 318–40.

Lewandowsky, S. (2016), "Future Global Change and Cognition," *Topics in Cognitive Science*, 8: 7–18.

Lewandowsky, S. (2021), "Climate Change Disinformation and How to Combat it," *Annual Review of Public Health*, 42: 1–21.

Lewandowsky, S., Cook, J., and Lloyd, E. (2018), "The 'Alice in Wonderland' Mechanics of the Rejection of (Climate) Science: Simulating Coherence by Conspiracism," *Synthese*, 195 (1): 175–96.

Lewandowsky, S., Ecker, U., and Cook, J. (2017), "Beyond Misinformation: Understanding and Coping with the Post-truth Rra," *Journal of Applied Research in Memory and Cognition,* 6 (4): 353–69.

Lewandowsky, S., Oberauer, K., and Gignac, G. E. (2013), "NASA Faked the Moon Landing—Therefore, (Climate) Science Is a Hoax: An Anatomy of the Motivated Rejection of Science," *Psychological Science,* 24 (5): 622–33.

Levy, N. (2007), "Radically Socialized Knowledge and Conspiracy Theories," *Episteme*, 4 (2): 181–92.

Livingstone, D. N. (2002a), *Science, Space and Hermeneutics: Hettner-Lecture 2001*, Heidelberg: Department of Geography, University of Heidelberg.

Livingstone, D. N. (2002b), "Ecology and the Environment," in G. B. Ferngren (ed.), *Science and Religion: A Historical Introduction*, 345–57. Baltimore: The John Hopkins University Press.

Livingstone, D. N. (2003), *Putting Science in its Place*, Chicago: University of Chicago Press.

Livingstone, D. N. (2005a), "Text, Talk and Testimony: Geographical Reflections on Scientific Habits. An Afterword," *The British Journal for the History of Science*, 38 (1): 93–100.

Livingstone, D. N. (2005b), "Science, Text and Space: Thoughts on the Geography of Reading," *Transactions of the Institute of British Geographers*, 30 (4): 391–401.

Lowry, E. (2016), "God's Greenhouse: Eco-Evangelicals, Rhetoric, and the Public Sphere," *CrossCurrents*, 66 (1): 57–69.

Lunenfeld, P. (1999), "Screen Grabs: The Digital Dialectic and New Media," in P. Lunenfeld (ed.), *The Digital Dialectic: New Essays on New Media*, xiv–xxi, Cambridge, MA: MIT Press.

Luton, L. S. (2015), "Climate Scientists and the Intergovernmental Panel on Climate Change: Evolving Dynamics of a Belief in Political Neutrality," *Administrative Theory & Praxis*, 37: 144–61.

Mahl, D., Schäfer, M. S., and Zeng, J. (2022), "Conspiracy Theories in Online Environments: An Interdisciplinary Literature Review and Agenda for Future Research," *New Media & Society* [online]. Available at: https://doi.org/10.1177/14614448221075759

Mahony, M. (2013), "Boundary Spaces: Science, Politics and the Epistemic Geographies of Climate Change in Copenhagen, 2009," *Geoforum*, 49: 29–39.

Mahony, M. and Hulme, M. (2018), "Epistemic Geographies of Climate Change: Science, Space and Politics," *Progress in Human Geography*, 42 (3): 395–424.

Manderson, L. and Levine, S. (2020), "COVID-19, Risk, Fear, and Fall-Out," *Medical Anthropology*, 39 (5): 367–70.

Maréchal, N. (2016), "Automation, Algorithms, and Politics| when bots Tweet: Toward a Normative Framework for Bots on Social Networking Sites (feature)," *International Journal of Communication*, 10 [online]. Available at: https://ijoc.org/index.php/ijoc/article/view/6180/1811

Margolis, M. F. (2020), "Who Wants to Make America Great Again? Understanding Evangelical Support for Donald Trump," *Politics and Religion*, 13(1), 89–118.

Marsden, G. (1980), *Fundamentalism and American Culture: The Shaping of Twentieth Century Evangelicalism, 1870-1925*, New York: Oxford University Press.

Marsden, G. (1991), *Understanding Fundamentalism and Evangelicalism*, Grand Rapids: Wm. B. Eerdmans Publishing Co.

Martin, C. (2017), *A Critical Introduction to the Study of Religion*, London: Routledge.

Massey, D. (1993), "Questions of Locality," *Geography*, 78 (2): 142–9.

McAlister, M. (2003), "Prophecy, Politics, and the Popular: The Left Behind Series and Christian Fundamentalism's New World Order," *The South Atlantic Quarterly*, 102 (4): 773–98.

McAlister, M. (2018), *The Kingdom of God Has No Borders: A Global History of American Evangelicals*, Oxford: Oxford University Press.

McAlister, C. (2020), "Through a Glass Darkly: Time, the End, and the Essence of Apocalyptica," in McAlister, C. (ed.), *The Cambridge Companion to Apocalyptic Literature*, 1–18, Cambridge: Cambridge University Press.

McCammack, B. (2007), "Hot Damned America: Evangelicalism and the Climate Change Policy Debate," *American Quarterly*, 59 (3): 645–68.

McClure, P. K. (2017), "Tinkering with Technology and Religion in the Digital Age: The Effects of Internet Use on Religious Belief, Behavior, and Belonging," *Journal for the Scientific Study of Religion*, 56 (3): 481–97.

McLean, J. (2020), *Changing Digital Geographies Technologies, Environments and People*, Cham: Springer Nature, Palgrave Macmillan.

Megoran, N., (2012), "Radical Politics and the Apocalypse: Activist Readings of Revelation," *Area*, 45 (2): 141–7.

Miller, S. P. (2014), *The Age of Evangelicalism*, New York: Oxford University Press.

Morris, D. S. and Morris, J. S. (2013), "Digital Inequality and Participation in the Political Process: Real or Imagined?" *Social Science Computer Review*, 31 (5): 589–600.

Moser, P. K. (2002), "Introduction," in P. K. Moser (ed.), *The Oxford Handbook of Epistemology*, 3–24, Oxford: Oxford University Press.

Muirhead, R. and Rosenblum, N. L. (2019), *A Lot of People Are Saying*, Princeton: Princeton University Press.

Müller, M. (2008), "Reconsidering the Concept of Discourse for the Field of Critical Geopolitics: Towards Discourse as Language and Practice," *Political Geography*, 27 (3): 322–38.

Noll, M. A. (2002), "Evangelicalism and Fundamentalism," in G. B. Ferngren (ed.), *Science & Religion*, 261–76, Baltimore: John Hopkins University Press.

Northcott, M. (2004), *An Angel Directs the Storm: Apocalyptic Religion and American Empire*, London: I.B. Tauris.

O'Donnell, J. (2019) "Unipolar Dispensations: Exceptionalism, Empire, and the End of One America," *Political Theology*, 20 (1): 66–84.

O'Donnell, J. (2020a), "The Deliverance of the Administrative State: Deep State Conspiracism, Charismatic Demonology, and the Post-Truth Politics of American Christian Nationalism," *Religion*, 50 (4): 696–719.

O'Donnell, J. (2020b), *Passing Orders: Demonology and Sovereignty in American Spiritual Warfare*, Fordham University Press.

O'Donnell, J. (2021), "Antisemitism Under Erasure: Christian Zionist Anti-Globalism and the Refusal of Cohabitation," *Ethnic and Racial Studies*, 44 (1): 39–57.

O'Donnell, J. (2022), "Damned Ecologies: Environmental Demonology and Apocalyptic Normativization in American Spiritual Warfare," *Environmental Humanities*, 14 (3): 543–63.

O'Leary, S. D. (1998), *Arguing the Apocalypse: A Theory of Millennial Rhetoric*, Oxford: Oxford University Press.

Ó Tuathail, G (1998), "Postmodern Geopolitics? The Modern Geopolitical Imagination and Beyond," in Ó Tuathail G. and S. Dalby(ed.), *Rethinking Geopolitics*, 16–38, London: Routledge.

Ó Tuathail, G. and Dalby, S. (1998), "Introduction: Rethinking Geopolitics: Towards a Critical Geopolitics," in Ó Tuathail G. and S. Dalby(ed.), *Rethinking Geopolitics*, 1–15, London: Routledge.

Opderbeck, D. (2022), "Donald Trump and the End Times: How Dispensational Premillennialism Connects Christians with the Big Election Lie," *University of St. Thomas Journal of Law and Public Policy*, 15 (2): 545–90.

Ophir, A. and Shapin, S. (1991), "The Place of Knowledge: A Methodological Survey," *Science in Context*, 4 (1): 3–21.

Oxford Dictionary. (2016), *Word of the Year 2016 is...*, Available at: https://en.oxforddictionaries.com/word-of-the-year/word-of-the-year-2016 (Last Accessed: September 22, 2017).

Paltridge, B. (2012), *Discourse Analysis*, 2nd edn, London and New York: Bloomsbury Academic.

Pasgaard, M., Dalsgaard, B., Maruyama, P. K., Sandel, B., and Strange, N. (2015), "Geographical Imbalances and Divides in the Scientific Production of Climate Change Knowledge," *Global Environmental Change*, 35: 279–88.

Pearse, R. (2017), "Gender and Climate Change," *WIREs Clim Change*, 8 (2017): 1-16.

Perry, S. (2024), "Why Evangelicals Went All In on Trump, Again," *Time*, (01/24). https://time.com/6588138/evangelicals-support-donald-trump-2024/

Persily, N. (2017), "The 2016 US Election: Can Democracy Survive the Internet?" *Journal of Democracy*, 28 (2): 63–76.

Peterson, A. (2000), "In and Of the World? Christian Theological Anthropology and Environmental Ethics," *Journal of Agricultural and Environmental Ethics*, 12 (3): 237–61.

Pew Research Center. (2015), *America's Changing Religious Landscape*. Available at: http://www.pewforum.org/religious-landscape-study/ (Last Accessed: April 14, 2017).

Pew Research Center. (2016), *Evangelicals Increasingly say it's Becoming Harder for them in America*, July 14. Available at: https://www.pewresearch.org/fact-tank/2016/07/14/evangelicals-increasingly-say-its-becoming-harder-for-them-in-america/ (Last Accessed: September 14, 2021).

Pew Research Center. (2019) *In U.S., Decline of Christianity Continues at Rapid Pace*, October 17. Available at: https://www.pewresearch.org/religion/2019/10/17/in-u-s-decline-of-christianity-continues-at-rapid-pace/ (Last Accessed: September 14, 2021).

Pinkerton, A. and Benwell, M. (2014), 'Rethinking Popular Geopolitics in the Falklands/Malvinas Sovereignty Dispute: Creative Diplomacy and Citizen Statecraft," *Political Geography*, 38: 12–22.

Pirro, A. L. and Taggart, P. (2022), "Populists in Power and Conspiracy Theories," *Party Politics* [online]. Available at: https://doi.org/10.1177/13540688221077071

Rathje, J. (2021), "For Reich and Volksgemeinschaft—Against the World Conspiracy: Antisemitism and Sovereignism in the Federal Republic of Germany Since 1945," *Antisemitism Studies*, 5 (1): 100–38.

Ravetz, J. R. (1986), "Usable Knowledge, Usable Ignorance: Incomplete Science with Policy Implications," in W. C. Clark and R. C. Munn (ed.), *Sustainable Development of the Biosphere*, 415–32, New York: Cambridge University Press.

Ravetz, J. R. (1999), "What is Post-Normal Science," *Futures*, 31: 647–53.

Ravetz, J. R. (2011), "'Climategate' and the Maturing of Post-Normal Science," *Futures*, 43 (2011): 149–57.

Rey, T. (2014), *Bourdieu on Religion Imposing Faith and Legitimacy*, London and New York: Routledge.

Ricker, A. (2020), "Crisis, Conspiracy, and Community in Evangelical Climate Denial," *Journal of the Council for Research on Religion*, 2 (1): 72–91.

Roberts, M. (2008), *Evangelicals and Science*, Westport, CT: Greenwood Press.
Roberts, M. (2012), "Evangelicals and Climate Change," in D. Gerten and S. Bergmann (ed.), *Religion in Environmental and Climate Change- Suffering, Values, Lifestyles*, 107–31, London: Continuum international Publishing Group.
Robertson, D. G. (2015), "Silver Bullets and Seed Banks: A Material Analysis of Conspiracist Millennialism," *Nova Religio: The Journal of Alternative and Emergent Religions*, 19 (2): 83–99.
Robertson, D. G. (2016), *UFOs, Conspiracy Theories and the New Age*, London: Bloomsbury.
Robertson, D. G. (2018), "The Counter-Elite: Strategies of Authority in Millennial Conspiracism' in A. Dyrendal, D. G. Robertson, and E. Asprem (ed.), *Handbook of Conspiracy Theory and Contemporary Religion*, 234–56, Leiden: Brill.
Robertson, D. G., Asprem, E., and Dyrendal, A. (2018), "Introducing the Field: Conspiracy Theory in, about, and as Religion," in A. Dyrendal, D. G. Robertson, and E. Asprem (ed.), *Handbook of Conspiracy Theory and Contemporary Religion*, 1–20, Leiden: Brill.
Röchert, D., Shahi, G. K., Neubaum, G., Ross, B., and Stieglitz, S. (2021), "The Networked Context of COVID-19 Misinformation: Informational Homogeneity on YouTube at the Beginning of the Pandemic," *Online Social Networks and Media*, 26 [online]. Available at: https://www.sciencedirect.com/science/article/pii/S246869642100046X
Rode, D. C. and Fischbeck, P. S. (2021), "Apocalypse now? Communicating Extreme Forecasts," *International Journal of Global Warming*, 23 (2): 191–211.
Rolin, K. (2007), "Science as Collective Knowledge," *Cognitive Systems Research*, 9 (2008): 115–24.
Ronan, M. (2017), "Religion and the Environment: Twenty-first Century American Evangelicalism and the Anthropocene," *Humanities*, 6 (4) [online]. Available at: https://doi.org/10.3390/h6040092
Rozario, K. (2001), "What Comes down must go up: Why Disasters have been Good for American Capitalism," in S. Biel (ed.), *American Disasters*, 72–102, New York University Press.
Roser-Renouf, C., Maibach, E., Leiserowitz, A., and Rosentahl, S. (2016), *Global Warming, God and the "End Times",* Yale Program on Climate Chane Communication, Climate Note, July 26. Available at: https://climatecommunication.yale.edu/publications/global-warming-god-end-times/
Rowlands, I. H. (2000), "Beauty and the Beast? BP's and Exxon's Positions on Global Climate Change," *Environment and Planning C: Government and Policy*, 18 (3): 339–54.
Schor, E. (2021), "Christianity on Display at Capitol Riot Sparks New Debate," *AP News*, January 28. Available at: https://apnews.com/article/christianity-capitol-riot-6f13ef0030ad7b5a6f37a1e3b7b4c898 (Last Accessed: October 29, 2022).
Schwadel, P. (2017), "The Republicanization of Evangelical Protestants in the United States: An Examination of the Sources of Political Realignment," *Social Science Research*, 62 (2017): 238–54.
Selby, J. and Hoffmann, C. (2014), "Rethinking Climate Change, Conflict and Security," *Geopolitics*, 19 (4): 747–56.
Shapin, S. (1988), "The House of Experiment in Seventeenth-Century England," *Isis*, 79 (3): 373–404.

Shapin, S. (1994), *A Social History of Truth*, Chicago, IL: The University of Chicago Press.

Shapin, S. (1998), "Placing the View from Nowhere: Historical and Sociological Problems in the Location of Science," *Transactions of the Institute of British Geographers*, 23 (1): 5–12.

Shapin, S. and Schaffer, S. (1985), *Leviathan and the Air-Pump: Hobbes, Boyle, and the Experimental Life*, Princeton, NJ: Princeton University Press.

Sharman, A. (2014), "Mapping the Climate Sceptical Blogosphere," *Global Environmental Change*, 26: 159–70.

Sharrock, W. and Read, R. (2002), *Kuhn: Philosopher of Scientific Revolution*, 4th edn, Malden, MA: Blackwell Publishing Group.

Sheldon, M. P. and Oreskes, N. (2017), "The Religious Politics of Scientific Doubt: Evangelical Christians and Environmentalism in the United States," in J. Hart (ed.), *The Wiley Blackwell companion to Religion and Ecology*, 348–67, Hoboken: Wiley Blackwell.

Sideris, L. H. (2006), "Religion, Environmentalism, and the Meaning of Ecology," in R. S. Gottlieb (ed.), *The Oxford Handbook of Religion and Ecology*, 446–66, Oxford and New York: Oxford University Press.

Sismondo, S. (2017), "Post-truth?" *Social Studies of Science*, 47 (1): 3–6.

Skrimshire, S. (2014), "Climate Change and Apocalyptic Faith," *Wiley Interdisciplinary Reviews: Climate Change*, 5 (2): 233–46.

Skrimshire, S. (2019), "Activism for End Times: Millenarian Belief in an Age of Climate Emergency," *Political Theology*, 20 (6): 518–36.

Snee, H., Hine, C., Morey, Y., Roberts, S., and Watson, H. (2016), "Digital Methods as Mainstream Methodology: An Introduction," in H. Snee, C. Hine, Y. Morey, S. Roberts, and H. Watson (eds.), *Digital Methods for Social Science*, 1–11, London: Palgrave Macmillan.

Soentgen, J. and Bilandzic, H. (2014), "Die Struktur klimaskeptischer Argumente: Verschwörungstheorie als Wissenschaftskritik," *GAIA*, 23 (1): 40–7.

Spark, A. (2001), "Conjuring Order: The New World Order and Conspiracy Theories of Globalization," *The Sociological Review*, 48 (2): 46–62.

Šrol, J., Ballová Mikušková, E., and Čavojová, V. (2021), "When We Are Worried, What Are We Thinking? Anxiety, Lack of Control, and Conspiracy Beliefs Amidst the COVID-19 Pandemic," *Applied Cognitive Psychology*, 35 (3): 720–9.

Steffaniak, J. L. (2020), "The God of All Creation: A Critique of Evangelical Biblicism and Recovery of Perfect Being Theology," *Journal of Reformed Theology*, 14 (4): 358–80.

Steinmetz-Jenkins, D. (2020), "The Nationalist Roots of White Evangelical Politics," *Dissent*, 67 (2): 27–31.

Stephens, R. J. and Giberson, K. W. (2011), *The Anointed: Evangelical Truth in a Secular Age*, Cambridge: Belknap Press.

Stewart, K. and Harding, S. (1999), "Bad Endings: American Apocalypsis," *Annual Review of Anthropology*, 28 (1999): 285–310.

Stover, D. (2019), "Evangelicals for Climate Action," *Bulletin of the Atomic Scientists*, 75(2), 66–72.

Stunt, T. C. F. (2012), "Trinity College, John Darby and the Powerscourt Milieu," in J. Searle and K. G. C. Newport (ed.), *The Future of Millennial Studies*, Sheffield: Sheffield Phoenix Press, 47–74.

Sturm, T. (2006), "Prophetic Eyes: The Theatricality of Mark Hitchcock's Premillennial Geopolitics," *Geopolitics,* 11 (2): 231–55.

Sturm, T. (2010), "Imagining Apocalyptic Geopolitics: American Evangelical Citationality of Evil Others," in J. Dittmer and T. Sturm, T. (eds.), *Mapping the End Times: American Evangelical Geopolitics and Apocalyptic Visions*, 133–54, Surrey: Ashgate Publishing Limited.

Sturm, T. (2012), "Imminent Immanence of Judeo-Evangelical Nationalism: American Christian Zionists and Israel as the Future Redeemer Nation and State," *Relegere: Studies in Religion and Reception*, 2 (2): 333–41.

Sturm, T. (2013), "The Future of Religious Geopolitics: Towards a Research and Theory Agenda," *Area*, 45 (2): 134–40.

Sturm, T. (2017), "Christian Zionism as Religious Nationalism Par Excellence," *Brown Journal of World Affairs*, 24 (1): 7–21.

Sturm, T. (2018), "Religion as Nationalism: The Religious Nationalism of American Christian Zionists," *National Identities*, 20 (39): 299–319.

Sturm, T. and Albrecht, T. (2021a), "Constituent Covid-19 Apocalypses: Contagious Conspiracism, 5G, and Viral Vaccinations," *Anthropology & Medicine*, 28 (1): 122–39.

Sturm, T. and Albrecht, T. (2021b), "Hal Lindsey," *Critical Dictionary of Apocalyptic and Millenarian Movements*, CenSAMM, Available at: https://www.cdamm.org/articles/hal-lindsey (Last Accessed: June 25, 2022).

Sturm, T. and Dittmer, J. (2010) "Introduction: Mapping the End Times," in J. Dittmer and T. Sturm (eds.) *Mapping the End Times: American Evangelical Geopolitics and Apocalyptic Visions*, 1–24, Surrey: Ashgate Publishing Limited.

Sturm, T. and Lustig, N. (2022), "Variegated Environmental Apocalypses: Post-Politics, the Contestatory, and an Eco-Precariat Manifesto for a Radical Apocalyptics," in E. T. Harper and D. Specht (eds.), *Imagining Apocalyptic Politics for the Anthropocene*, 213–34, Abingdon: Routledge.

Suhay, E. and Druckman, J. N. (2015), "The Politics of Science: Political Values and the Production, Communication, and Reception of Scientific Knowledge," *The ANNALS of the American Academy of Political and Social Science,* 658: 6–15.

Sutton, M. A. (2012), "Was FDR the Antichrist? Birth of Fundamentalist Antiliberalism in a Global Age," *The Journal of American History*, 1052–74.

Sutton, M. A. (2017), *American Apocalypse: A History of Modern Evangelicalism*, Cambridge and London: The Belknap Press of Harvard University Press.

Swami, V. and Furnham, A. (2014), "Political Paranoia and Conspiracy Theories," in J.-W. Van Prooijen and P. A. M. Van Lange (eds.), *Power Politics and Paranoia*, 218–36, Cambridge: Cambridge University Press.

Swartz, D. R. (2011), "Identity Politics and the Fragmenting of the 1970s Evangelical Left," *Religion and American Culture: A Journal of Interpretation*, 21 (1): 81–120.

Swartz, D. R. (2012), "The Evangelical Left and the Future of Social Conservatism," *Society,* 49 (1): 54–60.

Swyngedouw, E. (2010), "Apocalypse Forever?," *Theory, Culture & Society*, 27 (2–3): 213–32.

Swyngedouw, E. (2013), "Apocalypse Now! Fear and Doomsday Pleasures," *Capitalism Nature Socialism*, 24 (1): 9–18.

Taylor, B., Van Wieren, G., and Zaleha, B. (2016), The Greening of Religion Hypothesis (Part Two): Assessing the Data from Lynn White, Jr, to Pope Francis," *Journal for the Study of Religion, Nature and Culture*, 10 (3): 306–78.

Taylor, C. (2007), *A Secular Age*, Cambridge, MA: The Belknap Press of Harvard University Press.

Törnberg, P. (2018), "Echo Chambers and Viral Misinformation: Modeling Fake News as Complex Contagion," *PLoS One*, 13 (9) [online]. Available at: https://doi.org/10.1371/journal.pone.0203958

Tsuria, R., Yadlin-Segal, A., Vitullo, A., and Campbell, H. A. (2017), "Approaches to Digital Methods in Studies of Digital Religion," *The Communication Review*, 20 (2): 73–97.

Turnpenny, J. (2012), "Lessons from Post-Normal Science for Climate Science-Sceptic Debates," *WIREs Clim Change*, 2 (2012): 397–407.

Turnpenny, J., Jones, M., and Lorenzoni, I. (2011), "Where Now for Post-Normal Science?: A Critical Review of its Development, Definitions, and Uses," *Science, Technology, & Human Values*, 36 (3): 287–306.

Ueno, Y., Hyodo, M., Yang, T., and Katoh, S. (2019), "Intensified East Asian Winter Monsoon during the Last Geomagnetic Reversal Transition," *Scientific Reports*, 9 (1): 1–8.

Uscinski, J. E. (2020), *Conspiracy Theories: A Primer*, Lanham, MD: Rowman and Littlefield.

Uscinski, J. E. and Parent, J. M. (2014), *American Conspiracy Theories*, Oxford: Oxford University Press.

Uscinski, J. E., Klofstad, C., and Atkinson, M. D. (2016), "What Drives Conspiratorial Beliefs? The Role of Informational Cues and Predispositions," *Political Research Quarterly*, 69 (1): 57–71.

Uscinski, J. E., Douglas, K., and Lewandowsky, S. (2017), "Climate Change Conspiracy Theories," in *Oxford Research Encyclopedia of Climate Science* [online]. Available at: https://doi.org/10.1093/acrefore/9780190228620.013.328

USGS. (2015), *Coastal-change and Glaciological Map of the Amery Ice Shelf Area, Antarctica: 1961–2004*. Available at: https://pubs.er.usgs.gov/publication/i2600Q (Accessed: September 10, 2020).

Van den Bulck, H. and Hyzen, A. (2020), "Of Lizards and Ideological Entrepreneurs: Alex Jones and Infowars in the Relationship between Populist Nationalism and the Post-Global Media Ecology," *International Communication Gazette*, 82 (1): 42–59.

Van der Linden, S., Leiserowitz, A., Rosenthal, S., and Maibach, E. (2017), "Inoculating the Public Against Misinformation about Climate Change," *Global Challenges*, 1 (2) [online]. Available at: https://doi.org/10.1002/gch2.201600008

van Prooijen, J. (2020), "An Existential Threat Model of Conspiracy Theories," *European Psychologist*, 25 (1): 16–25.

van Prooijen, J. and Douglas, K. M. (2017), "Conspiracy Theories as Part of History: The Role of Societal Crisis Situations," *Memory Studies*, 10 (3): 323–33.

Veldman, R. G. (2019), *The Gospel of Climate Skepticism: Why Evangelical Christians Oppose Action on Climate Change*, Berkeley: University of California Press.

Verter, B. (2003), 'Spiritual Capital: Theorizing Religion with Bourdieu Against Bourdieu," *Sociological Theory*, 21 (2): 150–74.

Waldman, S. (2024), "No More Going Wobbly in Climate Fight, Trump Supporters Vow," *Politico*, (1/16). https://www.politico.com/news/2024/01/12/trump-second-term-climate-science-2024-00132289

Ward, C. and Voas, D. (2011), "The Emergence of Conspirituality," *Journal of Contemporary Religion*, 26 (1): 103–21.

Watts, J. (2018), "We have 12 Years to Limit Climate Change Catastrophe, Warns UN," *The Guardian*, October 8. Available at: https://www.theguardian.com/environment/2018/oct/08/global-warming-must-not-exceed-15c-warns-landmark-un-report (Last Accessed: June 25, 2022).

Webster, J. (2021), "Embodied Apocalypse: Or the Native Cosmology of Late Modern Social Theory," *Anthropology & Medicine*, 28 (1): 13–27.

Weingart, P. (1997), "From 'Finalization' to 'Mode 2': Old Wine in New Bottles," *Social Science Information*, 36: 591–613.

Wessinger, C. (1997), "Millennialism With and Without the Mayhem," in T. Robbins and S. J. Palmer (eds.), *Millennium, Messiahs, and Mayhem: Contemporary Apocalyptic Movements*, 47–60, New York: Routledge.

Werly, J. M. (1977), "Premillennialism and the Paranoid Style," *American Studies*, 18 (1): 39–55.

White Jr., L. (1967), "The Historical Roots of Our Ecologic Crisis," *Science*, 155 (3767): 1203–7.

Whitehead, A. L. and Perry, P. L. (2020), *Taking America Back for God: Christian Nationalism in the United States*, Oxford: Oxford University Press.

Williams, E. C. (2010), *Combined and Uneven Apocalypse: Luciferian Marxism*, Ropley: John Hunt Publishing/Zero Books.

Wilson, A. F. (2017), "The Bitter End: Apocalypse and Conspiracy in White Nationalist Responses to the Islamic State Attacks in Paris," *Patterns of Prejudice*, 51 (5): 412–31.

Wong, J. (2015), "The Role of Born-Again Identity on the Political Attitudes of Whites, Blacks, Latinos, and Asian Americans," *Politics and Religion*, 8 (4): 641–78.

Wong, J. (2018a), "The Evangelical Vote and Race in the 2016 Presidential Election," *Journal of Race, Ethnicity, and Politics*, 3 (1): 81–106.

Wong, J. (2018b), *Immigrants, Evangelicals, and Politics in an Era of Demographic Change*, New York: Russell Sage Foundation.

Wood, M. and Douglas, D. (2018), "Are Conspiracy Theories a Surrogate for God?' in A. Dyrendal, D. G. Robertson, and E. Asprem (eds.), *Handbook of Conspiracy Theory and Contemporary Religion*, 87–105, Leiden: Brill.

Worthen, M. (2014), *Apostles of Reason*, New York: Oxford University Press.

Wynne, B. (2010), "Strange Weather, Again," *Theory, Culture & Society*, 27 (2–3): 289–305.

YouTube Help. (no date), *See Fact Checks in YouTube Search Results*. Available at: https://support.google.com/youtube/answer/9229632?hl=en (Last Accessed: June 26, 2022).

Zaleha, B. D. and Szasz, A. (2014), "Keep Christianity Brown! Climate Denial on the Christian Right in the United States," in G. R. Veldman, A. Szasz, and R. Haluza-DeLay (eds.), *How the World's Religions Are Responding to Climate Change*, 209–28, New York: Routledge.

Zaleha, B. D. and Szasz, A. (2015), "Why Conservative Christians Don't Believe in Climate Change," *Bulletin of the Atomic Scientists*, 71 (5): 19–30.

# Primary Sources

Becker, M. A. (2021), "Climate and Prophecy," *RaptureReady*, October 10. Available at: https://www.raptureready.com/2021/10/07/climate-and-prophecy-by-mark-a-becker/ (Last Accessed: April 28, 2022).

Beisner, C. (2015), "32 Books I Read in 2015 (Make that 33)," *Cornwall Alliance*, December 21. Available at: https://cornwallalliance.org/2015/12/32-books-i-read-in-2015/ (Last Accessed: April 28, 2022).

Beisner, C. (2020a). "So, We've Rejected Science, Have We?" *Cornwall Alliance*, April 1. Available at: https://cornwallalliance.org/2020/04/so-weve-rejected-science-have-we/ (Last Accessed: April 28, 2022).

Brentner, J. C. (2019), "Heartbreak," *RaptureReady*, October 10. Available at: https://www.raptureready.com/2019/10/12/heartbreak-jonathan-brentner/ (Last Accessed: May 10, 2022).

Brentner, J. C. (2020), "COVID-19: Transition to the New World Order," *RaptureForums*, May 15. Available at: https://www.raptureforums.com/end-times/covid-19-transition-to-the-new-world-order/ (Last Accessed: May 10, 2022).

Burnett, H. S. (2021), "Climate Alarmists Call for Global 'Eco-Dictatorship,'" *Cornwall Alliance*, August 21. Available at: https://cornwallalliance.org/2021/08/climate-alarmists-call-for-global-eco-dictatorship/ (Last Accessed: April 28, 2022).

Davidson, G. (2020), "Climate Change: Truth or Fiction," *Rapture Ready*, March 3. Available at: https://www.raptureready.com/2020/03/06/climate-change-truth-or-fiction-by-gale-davidson/ (Last Accessed: June 25, 2022).

Duck, D. (2018), "NWO and Quotes," originally posted on *RaptureReady*, August 5. Available at: https://www.blogarama.com/religion-blogs/1502-are-living-end-times-blog/27598678-nwo-quotes (Last Accessed: June 25, 2022)

Duck, D. (2020), "Enemies of God," *RaptureReady*, April 19. Available at: https://www.raptureready.com/2020/04/19/enemies-of-god-by-daymond-duck/ (Last Accessed: May 10, 2022).

Durden, T. (2019), "Bombshell Claim: Scientists Find 'Man-made Climate Change Doesn't Exist In Practice',' *ZeroHedge*, July 12. Available at: https://www.zerohedge.com/news/2019-07-11/scientists-finland-japan-man-made-climate-change-doesnt-exist-practice

Ehret, M. (2021), "Exposing the Great Reset and Green New Deal Frauds," *Z3 News*, August 25. Available at: https://z3news.com/w/exposing-great-reset-green-new-deal-frauds/ (Last Accessed: June 25, 2022).

EndTimeHeadlines. (2021), "Bill Gates Unveils His Master Plan for Battling 'Climate Change'...," *End Time Headlines*, February 15. Available at: https://endtimeheadlines.org/2021/02/bill-gates-unveils-his-master-plan-for-battling-climate-change/ (Last Accessed: June 25, 2022).

EndTimeInc. (2017), "The Global Warming Religion | Endtime Ministries with Irvin Baxter," *YouTube*, January 10. Available at: https://www.youtube.com/watch?v=W-m8UW6XFC0&t=1s (Last Accessed: June 25, 2022).

EndTimeInc. (2019), "The Climate Change Hoax," [online video]. *YouTube*, September 5. Available at: https://www.youtube.com/watch?v=RuiUFegqR7Q (Last Accessed: June 25, 2022).

EndTimeMinistries. (2019a), "Italy's Leftist Government to Indoctrinate Schoolchildren in 'Climate Change',' *EndTimeMinistries*, November 7.

Available at: https://www.endtime.com/prophecy-news/italys-leftist-government-to-indoctrinate-schoolchildren-in-climate-change/ (Last Accessed: June 25, 2022).

EndTimeMinistries. (2019b), "Major Newspaper will refer to Climate Change as an 'Emergency,' *'Breakdown'*," *EndTimeMinistries*, May 24. Available at: https://www.endtime.com/prophecy-news/major-newspaper-will-refer-to-climate-change-as-an-emergency-breakdown/ (Last Accessed: June 25, 2022).

EndTimesBibleProphecy. (no date), "The Unhindered March Toward Globalism," *End Times Bible Prophecy*. Available at: http://www.end-times-bible-prophecy.com/the-unhindered-march-toward-globalism.html (Last Accessed: April 28, 2022).

FaithWriters. (no date), "Mark A. Becker," *FaithWriters*. Available at: https://www.faithwriters.com/member-profile.php?id=38120 (Last Accessed: June 25, 2022).

FeedSpot Blog. (2018-2022a), "35 Best End Times Blogs and Websites To Follow in 2022," Available at: https://blog.feedspot.com/end_times_blogs/ (Last Accessed: December 29, 2022).

FeedSpot Blog. (2018-2022b), "70 Best Prophecy Blogs and Websites," Available at: https://blog.feedspot.com/prophecy_blogs/?_src=search (Last Accessed: December 29, 2022)

Gal, E. (2020a), "Pope Francis Says New World Order Needs to Happen Now with United Nations in Charge & Pope Francis Endorses Coronavirus 'Vaccines For All'," *Z3 News*, October 12. Available at: https://z3news.com/w/pope-francis-world-order-happen-united-nations-charge/ (Last Accessed: April 18, 2022).

Gal, E. (2020b), "The Plan Revealed – A Permanent 'Climate Lockdown' Because It Is Good For The Earth Contrasted With Scripture With Mankind Given Dominion Over The Earth," *Z3 News*, September 26. Available at: https://z3news.com/w/plan-revealed-permanent-lockdowns-good-earth/ (Last Accessed: June 25, 2022).

Gillette, B. (2019), "The Unhindered March Toward Globalism," *RaptureReady*, June 20, Available at: https://www.raptureready.com/2019/06/20/march-toward-globalism/ (Last Accessed: June 25, 2022).

Greenfield, D. (2017), "The Climate Confederacy," *Rapture Forums*. Available at: https://www.raptureforums.com/politics-culture-wars/the-climate-confederacy (Last Accessed: June 25, 2022).

Hescox, M. and Ball, J. (2017), "CLIMATE LEADERSHIP IS NOT ABOUT PRESIDENT TRUMP: IT'S ABOUT YOU; IT'S ABOUT US," Available at: *http://www.creationcare.org/climate_leadership_is_not_about_president_trump* (Last Accessed: July 15, 2017).

Hescox, M. and Douglas, P. (2016), *Caring for Creation: The Evangelical's Guide to Climate Change and a Healthy Environment*, Bloomington: Baker Publishing Group.

Infowars. (2020), "Tin Foil Hat: Alex Jones Discusses Coronavirus, Censorship & Total Surveillance," *Infowars*. Available at: https://web.archive.org/web/20200329021617/https://www.infowars.com/tin-foil-hat-alex-jones-discusses-coronavirus-censorship-total-surveillance/ (Last Accessed: June 21, 2022).

Infowars. (2021), "Propaganda: Past Presidents Push Experimental Covid Vaccines on Public," *Infowars*, March 11, Available at: https://www.infowars

.com/posts/propaganda-past-presidents-push-experimental-covid-vaccines-on-public/ (Last Accessed: June 21, 2022).

Infowars (2022), "Planned Global Collapse has Begun; Food Crisis from Lebanon to Peru Leads to Riots in the Streets: Live Shows 4/11/22," *Infowars*, April 11. Available at: https://www.infowars.com/posts/biden-goes-after-guns-in-war-against-second-amendment-monday-live/ (Last Accessed: April 12, 2022).

James, T. (2021), "Wicked Worship Forewarned," *Rapture Forums*, July 12. Available at: https://www.raptureforums.com/end-times/wicked-worship-forewarned/ (Last Accessed: June 25, 2022).

Kinley, J., Moore, T. and Jones, N. (2021), "Interview with the Antichrist," *Christ in Prophecy / Lamb & Lion Ministries Talkshow*, February 28. Available at: https://christinprophecy.org/sermons/interview-with-the-antichrist/ (Last Accessed: May 10, 2022).

Kinsella, J. and Garcia, P. (2010), "Confessions of a Flat-Earther," *Omega Letters*, September 29. Available at: https://omegalettercom.wordpress.com/2010/09/29/confessions-of-a-flat-earther/ (Last Accessed: June 25, 2022).

Kuhnley, T. (2020), "How Christians Should Vote with Marsha Kuhnley," *Lamb & Lion Ministries*, October 25. Available at: https://christinprophecy.org/sermons/how-christians-should-vote-with-marsha-kuhnley/ (Last Accessed: June 25, 2022).

Lamb & Lion Ministries (2016a), "Lamplighter: The Pre-Tribulation Rapture: A Myth or a Reality?" Lamb & Lion Ministries, January/February Issue. Available at: http://lamblion.com/xfiles/publications/magazines/Lamplighter_JanFeb16_Rapture-Defense.pdf (Last Accessed: June 25, 2022).

Lamb & Lion Ministries (2016b), "Interviews of Woods and Batista," *Christ in Prophecy / Lamb & Lion Ministries*, February 21. Available at: https://christinprophecy.org/sermons/interviews-of-woods-and-batista/ (Last Accessed: April 28, 2022).

Lamb & Lion Ministries (2018), "Lamplighter: The Signs of Nature," *Lamb & Lion Ministries*, May/June Issue. Available at: https://lamblion.com/xfiles/publications/magazines/Lamplighter_MayJun20_Signs-of-Nature.pdf.

Lamb & Lion Ministries (2020), "Lamplighter: Pandemic," *Lamb & Lion Ministries*, April Special Edition, Available at: https://lamblion.com/xfiles/publications/magazines/Lamplighter_Apr20_Pandemic.pdf (Last Accessed: June 25, 2022).

Lamb & Lion Ministries (2022), "The Great Reset Revealed," *Christ in Prophecy / Lamb & Lion Ministries*, January 9. Available at: https://christinprophecy.org/sermons/the-great-reset-revealed/ (Last Accessed: June 25, 2022).

Law, S. (2018), "The Jesus Factor," *RaptureReady*, October 15. Available at: https://www.raptureready.com/2018/10/25/climate-jesus-factor-sally-law/ (Last Accessed: June 25, 2022).

Lindsey, H. (1980 [1970]), *The Late Great Planet Earth*, Eighty-first printing, Grand Rapids: Zondervan Publishing House.

Lindsey, H. (2015a), "Hal Lindsey Report: 2/27/2015," *The Hal Lindsey Report Website*. Available at: *http://www.hallindsey.com/videos/hal-lindsey-report-2272015/414* (Accessed: May 15, 2022).

Lindsey, H. (2015b), "Tonight on The Hal Lindsey Report," *The Hal Lindsey Report Website*, February 27. Available at: https://www.hallindsey.com/hlr-2-27-2015/ (Last Accessed: June 25, 2022).

REFERENCES

Lindsey, H. (2017), "Globalists and Climate Accords," *The Hal Lindsey Report Website*, June 3. Available at: https://www.hallindsey.com/ww-6-3-2017/ (Available at: https://www.hallindsey.com/ww-6-3-2017/ (Last Accessed: June 25, 2022).

Lindsey, H. (2019), "Special Announcement," *The Hal Lindsey Report*, June 10. Available at: https://www.hallindsey.com/ww-6-10-2019/ (Last Accessed: June 25, 2022).

Lindsey, H. (2022), "A Prophetic Preview," *The Hal Lindsey Report Website*, January 2. Available at: https://www.hallindsey.com/ww-1-2-2022/ (Last Accessed: June 25).

Martin, B. (2020), "Texas Anti-Vaxxers Fear Mandatory COVID-19 Vaccines More Than the Virus Itself," *Texas Monthly*, March 18. Available at: https://www.texasmonthly.com/news/texas-anti-vaxxers-fear-mandatory-coronavirus-vaccines/ (Last Accessed: June 25).

National Association of Evangelicals. (2017a), "What is an Evangelical?" Available at: https://www.nae.net/what-is-an-evangelical/ (Last Accessed: April 14, 2017).

Payne, D. (2016), "Is Climate Change Caused by the Power of Men?" *RaptureReady*, October 7. Available at: https://www.raptureready.com/2016/10/07/is-climate-change-caused-by-the-power-of-men-by-dan-payne/ (Last Accessed: June 25, 2022).

ProphecyNewsWatch. (2019), "The Religion Of Climate Change & The New Doomsday Scenario," *Prophecy News Watch*, September 26. Available at: https://z3news.com/w/plan-revealed-permanent-lockdowns-good-earth/ (Last Accessed:June 25, 2022).

RaptureForums. (2020), "Proposed CA Legislation Would Mandate Climate Change Indoctrination in Public Schools," *RaptureForums*. January 31. Available at: https://www.raptureforums.com/politics-culture-wars/proposed-ca-legislation-would-mandate-climate-change-indoctrination-in-public-schools/ (Last Accessed: June 25, 2022).

RaptureReady. (no date), "Timelines: Global Warming," *RaptureReady*. Available at: https://www.raptureready.com/timelines-global-warming/ (Last Accessed: June 25, 2022).

RaptureReady. (2021), "18 Oct 2021," *Rapture Ready*, Oct 18. Available at: https://www.raptureready.com/2021/10/17/pc-18-oct-2021/ (Last Accessed: June 25, 2022).

RaptureReady. (2022), "23 Apr 2022," *Rapture Ready*, April 23. Available at: https://www.raptureready.com/2022/04/23/23-apr-2022/ (Last Accessed: June 25, 2022).

Reese, G. (2020a), "5G Launches In Wuhan Weeks Before Coronavirus Outbreak," *Infowars*, January 30. Available at: https://www.infowars.com/5g-launches-in-wuhan-weeks-before-coronavirus-outbreak/#inline-comments.

Reese, G. (2020b), "666 The Mark Of The Beast Has Arrived," *Infowars*, April 20. Available at: https://www.infowars.com/666-the-mark-of-the-beast-has-arrived/.

Robbins, D. (2015), "Persuading the World to Worship the Antichrist, The Antichrist," *EndTimeMinistries*, November 12. Available at: https://www.endtime.com/articles-endtime-magazine/persuading-world-worship-antichrist/ (Last Accessed: April 18, 2022).

Robbins, D. (2019a), "Global Warming/Climate Change Scam," *EndTimeMinistries*, January 1. Available at: *https://www.endtime.com/end-of-the-age/global-warming-climate-change-scam-2/* (Last Accessed: May 15, 2022).

Robbins, D. (2019b), "Global Warming/Climate Change... The Deception," *EndTimeMinistries*, September 19. Available at: https://www.endtime.com/end-of-the-age/global-warming-climate-change-the-deception/ (Last Accessed: May 15, 2022).

Scaparo, R. (2021), "Google and YouTube will now Remove Monetization and restrict Ads from Content that Questions Global Warming," *End Time Headlines*, October 8 Available at: https://endtimeheadlines.org/2021/10/google-and-youtube-will-now-remove-monetization-and-restrict-ads-from-content-that-questions-global-warming/ (Last Accessed: May 15, 2022).

Schmutzer, S. (2017), "Global Warming," *Rapture Ready*, original now deleted. Copy available at: https://yesterdaysprophecy.com/global-warming-by-steve-schmutzer/ (Last Accessed: June 25, 2022).

SignsOfTheEndTimes. (no date), *Signs of the End Times*. Available at: https://www.signs-of-end-times.com/un-2030-jesus.html (Last Accessed: June 25, 2022).

Snyder, M. (2021), "Google and YouTube will now Remove Monetization and Restrict Ads from Content that Questions Global Warming," *End Time Headlines*, October 8. Available at: https://endtimeheadlines.org/2021/10/google-and-youtube-will-now-remove-monetization-and-restrict-ads-from-content-that-questions-global-warming/ (Last Accessed: June 25, 2022).

Stetzer, E. (2017), "Facts Are Our Friends: Why Sharing Fake News Makes Us Look Stupid and Harms Our Witness," *Christianity Today*, February 9. Available at: https://churchleaders.com/outreach-missions/outreach-missions-articles/299041-facts-friends-sharing-fake-news-makes-us-look-stupid-harms-witness.html (Last Accessed: December 28, 2022).

Strandberg, T. (2005), "Rapture Ready?" Raptureready.com (Last Accessed: Dec 2, 2005).

Strandberg, T. (2017), "Dodging the Climate Change Bullet," *RaptureForums*, June 11. Available at: https://www.raptureforums.com/politics-culture-wars/dodging-climate-change-bullet/ (Last Accessed: June 25, 2022).

Tomczak, L. (2019), "The Devil is Using the Democratic Party to Destroy America," *RaptureForums*, January 21. Available at: https://www.raptureforums.com/politics-culture-wars/the-devil-is-using-the-democratic-party-to-destroy-america/ (Last Accessed: April 28, 2022).

Truth Social. (2022), *Frontpage*. Available at: https://truthsocial.com/ (Last Accessed: June 25, 2022).

Ungurean, G. (2015), "Obama is Neither Inept Nor Inexperienced," *RaptureReady*, October 10. Available at: https://www.raptureready.com/2015/10/10/obama-neither-inept-inexperienced-geri-ungurean/ (Last Accessed: June 25, 2022).

Ungurean, G. (2016), "Preparing the World for Antichrist – The U.N. Sustainable Development Goals," *Rapture Ready*, August 30. Available at: https://www.raptureready.com/2016/08/30/preparing-the-world-for-antichrist-the-u-n-sustainable-development-goals-by-geri-ungurean/ (Last Accessed: September 19, 2017).

Ungurean, G. (2019) "The Lie of Man-Made Global Warming," *RaptureReady*, July 16. Available at: https://www.raptureready.com/2019/07/16/lie-man-made-global-warming-geri-ungurean/ (Last Accessed: June 25, 2022).

Veldman, R. G. (2019), *The Gospel of Climate Skepticism: Why Evangelical Christians Oppose Action on Climate Change*, Berkeley, CA: University of California Press.

White House. (2017), "Statement by President Trump on the Paris Climate Accord," June 1. Available at: https://it.usembassy.gov/statement-president-trump-paris-climate-accord/ (Last Accessed: June 25, 2022).

Woods, A. and McGowan, J. (2017), "2017.06.02. Prophetic Significance of US Withdrawal from Paris Climate Agreement." Available at: https://www.youtube.com/watch?v=fbAUry0oIyc&index=3&list=LLNVIraQwMZmf9IorMxHorag&t=3311s (Last Accessed: June 28, 2022).

World Economic Forum. (2020), *Greta Thunberg on Averting a Climate Apocalypse | DAVOS 2020,* [online video]. Available at: https://www.youtube.com/watch?v=51u4JECraLQ (Last Accessed: June 25, 2022).

Zitelmann, R. (2021), "To anti-Capitalists Climate Change is just a Pretext for a Planned Economy," *Cornwall Alliance*, August 26. Available at: https://cornwallalliance.org/2021/08/to-anti-capitalists-climate-change-is-just-a-pretext-for-a-planned-economy/ (Last Accessed: April 28, 2022).

Index

4chan 46, 51
5G 8
8chan 46, 51
9/11 81

ACC-skeptics 67, 90, 111, 116, 124
activists 1–2, 80, 114
Adamic 50
African Americans 64
Agenda 21 7, 94, 141, 143–4, 156
Agenda 30 94, 156
Agenda 21/30 7, 141, 143–4
alternative truths 6, 48
alt-right 126–7, 144, 152, 154
America First 18, 138, 146
American 2–6, 8–10, 12–19, 24–5, 29, 32–3, 36, 42, 47–8, 51–7, 59–79, 81–3, 85–9, 91, 93–7, 101–2, 106–8, 115, 117–18, 121–4, 128, 133, 135, 137–56, 158, 160–1
American Civil War 72, 85
American presidential election 8, 124
American right 4–6, 10, 12–14, 18, 78, 101, 148, 152, 158
anthropocentric 65, 68, 124
anthropogenic 4, 6, 14, 16, 18, 29–32, 34–5, 37–8, 41–3, 66–8, 86–9, 91, 105–13, 115–16, 121–2, 124–8, 132–4, 142, 150, 153, 160, 162
Antichrist 3, 5, 11–12, 16, 18, 56, 59, 69, 71–3, 75–82, 86–8, 91, 93–4, 96–8, 100–1, 105, 112, 114–15, 117, 137–45, 148–51, 153
anti-Christian 3, 16, 18, 42, 76–8, 80, 95, 97, 102, 105, 108, 111–12, 114, 119, 143, 145, 148–9

anti-globalism 3, 15, 118, 138–40, 144–6, 148, 150, 153
antiglobalist 81, 140, 142, 148, 152–3
anti-Soviet 76
antivaxxer 13
apocalypse 1–3, 9–12, 22, 54, 69, 71–2, 74, 78, 85–6, 89, 92, 97, 104, 115, 138, 142, 158–60
apocalyptic conspiracism 1–19, 22, 24, 26, 28, 30–2, 34, 36, 38, 40, 42, 44, 46, 48, 50, 52, 54, 56, 59–60, 62, 64, 66, 68, 70, 72–4, 76–83, 86–92, 94–6, 98–100, 102–6, 108–18, 122, 124, 126, 128, 130, 132, 134, 138, 140, 142, 144, 146, 148, 150, 152–4, 156, 158, 160–2
Apocalyptic Geopolitics 137, 139, 141, 143, 145, 147, 149, 151, 153
apocalyptic imaginations 1, 76, 87, 101
apocalypticism 1–2, 4–6, 8–10, 12, 15, 17–18, 32, 52–3, 55, 62, 73–4, 77–8, 80, 86, 91–3, 95–6, 98, 101–2, 104–5, 107–8, 118, 132, 135, 137, 141, 145, 149, 154, 158, 160–1
Armageddon 85, 96
Asad, Talal 4, 17, 61–2, 71, 139
astrology 54

Bacon, Francis 67
  Baconian 67–8, 90, 130, 132, 135
Barkun, Michael 8, 155
Bebbington, David W. 60
Bebbington Quadrilateral 60–2, 122
Bible Belt 63

biblical   1, 3, 54, 61, 65–8, 71–3, 78–80, 95, 99, 106, 121–5, 134, 137–8, 140–4, 148–9, 153, 159
Biblicism   54, 60–1, 68, 83, 122–3, 135
Biden, Joe   14, 32, 94, 153
biodiversity   34
Black   12, 14, 78, 88
black helicopters   78
Bolin, Bert   36
book of Revelation   1, 9, 141, 158
Bourdieu, Pierre   116
Breitbart   47, 112
Bush, President   19

Campbell, Heidi A.   52
Capitol Building   4, 48, 154
catastrophic millennialism   10
Catholic   66–7, 92, 99, 101–2
Catholicism   77, 101–2
charismatic   5, 59, 61, 80
Chinese   32
Christianity Today   75
Christian nationalism   3–4, 63, 152
Church Age   72, 80, 92, 97–8
citationality   6, 125, 141–2
civic epistemologies   30
civil rights   64
climate change   1–4, 6, 8, 10–11, 13–19, 21–2, 25, 27–44, 51–2, 54–7, 59–60, 62, 65–9, 74, 78–9, 81, 83, 85–94, 97–119, 121–35, 137–45, 147–53, 156–62
Climate Crisis   2–3, 94, 114–15
Climate Emergency   114–16
Climategate   111–12, 128
climate hazards   2
Clinton, Hillary   12, 47
Club of Rome   2
Cold War   76, 78, 85, 94, 139, 143, 148–9
collapse of human civilization   2
colonialism   14, 65
communism   76–7, 93, 97, 101–2
communist   93, 97, 102, 144, 152
conservatism   3, 63, 89
conspiracist   3–9, 11–18, 22–3, 25–7, 29–33, 35, 42–4, 48–52, 54–7, 59–60, 67, 69, 74–6, 78, 80–3, 85–102, 104–19, 121, 123–5, 127–8, 132, 138–9, 141–3, 147, 150–8, 160–2
conspiracy theorist   5, 51, 101, 109, 116
contagious conspiracism   43
COP21   86, 143
Cornwall Alliance   67–8, 89–92, 124, 128–9
counter-epistemic   31–2, 81, 117, 121–2, 135, 162
counter-hegemonic   31, 48, 74–5, 81–2, 104
counter-knowledge   13, 16–17, 21, 23, 25, 27, 29, 31–3, 35, 37, 39, 41–4, 48–52, 54, 56–7, 104–6, 108–9, 121, 123, 125, 127, 129, 131, 133–5, 154, 156
counter-truth   13, 155
Coverley, Desmond Michael   88
Covid   2–3, 8, 11, 13, 15, 22, 30, 32, 43, 51–2, 57, 75, 78, 81, 95, 111, 119, 141, 143–4, 156
Creation Care   66, 68
creationism   134
critical discourse analysis   44, 54–5, 116–17
cultural geographies   45, 95

Dallas Theological Seminary   63
Daniel, Book of   72, 76, 139–41
Darby, John Nelson   72, 97, 139
decarbonization   15
decarbonize   94
Deep State   14, 33, 42, 77, 148
degradation   1, 3, 65–6, 71, 88, 158
Democratic Party   11–12, 14, 32, 94, 138
Democrats   94, 151–2
demon   150, 152, 154
   demonic   98, 139, 151, 153
demonization   12, 150, 152–3
denialism   6, 14, 17, 162
determinism   149
Devil   94, 140
DiCaprio, Leonardo   10, 133
dictatorship   3, 78–9, 86, 90–1, 105, 113, 143–4
digital geographies   15, 17, 44, 47, 49

INDEX

digital religion 52
digital spaces 4, 6, 10, 15–19, 44, 46–53, 56–7, 91, 93, 95, 98, 102, 104, 109, 115, 118, 121–2, 124–5, 141, 149, 154–5
dispensationalism 69, 71–3, 80, 143
dispensational premillennialism 5, 17, 71, 73
distrust 28–9, 57, 147
Don't Look Up (film) 10
doomsday 1, 99
Doomsday Clock 1, 99
dystopian 1

earther 129
Earth Summit 98, 149
echo chamber 49–50
eco-apocalypse 158
eco-apocalypticism 1, 18, 74
eco-dictatorship 90–1
ecodictatorship 91
ecoevangelical 66
ecosystems 2, 34, 39
Ehrlich, Paul 2
elect 7, 12, 32, 75, 98, 138, 145, 156, 161
EndTimeHeadlines 56, 118–19, 142, 144
EndTimeMinistries 51, 100, 106, 110, 112, 114, 123, 131, 140–1, 151
End-Times 2–7, 12–14, 16–18, 53–6, 59, 71, 73–5, 78–9, 82, 85–8, 91, 95–6, 99, 102, 104, 106, 108, 110, 119, 121–2, 125–6, 128, 130, 132, 134, 137–8, 140, 142, 144–5, 147–51, 153–4, 156–9, 161
environmental 1–3, 17–18, 33, 35–41, 64–73, 80, 83, 86–94, 98–100, 104, 112, 115, 121–3, 126, 137–8, 158–62
environmental resources 2
epistemes 25–6, 31, 41, 45, 47
epistemic 15, 18, 21, 23, 25–39, 41–2, 44–8, 50, 53–4, 56–7, 67, 69, 81–2, 87, 90, 104, 106–7, 109, 116–17, 121–2, 125–6, 130, 132, 135, 149, 155–7, 160–2

epistemic authorities 21, 26, 28, 30–1, 33, 41, 44, 47, 50, 53, 57, 82, 90, 106–7, 109, 116, 121, 125–6, 126, 132, 135, 157, 160
epistemic capital 27, 29–30, 33–4, 42, 48
epistemic disposition 35, 47
epistemic geographies 35–6
epistemic Manicheanism 156
epistemic obligations 45
epistemic practice 18, 23, 25, 117, 155
epistemic subjects 23
epistemology 16, 19, 21–5, 27–31, 33, 35, 37, 39, 41, 44, 47, 81, 108, 117
eschatological 5, 61, 69–71, 73, 83, 85–6, 91, 96, 142–3, 146, 148
eschatology 3, 53, 63, 69–73, 76–7, 82, 85, 91, 97, 115, 129, 138–9, 145
esotericism 54
European Economic Community (EEC) 76–7, 139, 143
European Union 101, 137, 139
evangelical 2–6, 9–10, 13–15, 17–18, 30, 32, 47, 52–6, 59–71, 73–9, 81–3, 85–93, 95–102, 104–11, 113, 115–18, 121–35, 137–55, 157–8, 160–2
evangelical apocalyptic conspiracist 5, 17–18, 30, 56, 74, 78, 81–3, 85, 87, 92–3, 95, 97–8, 100–2, 104–9, 111, 115, 117–18, 121, 123–5, 151
Evangelical Environmental Network (EEN) 66, 68, 70–1, 132
evangelicalism 2, 4–5, 15, 17, 52–4, 59–67, 69–71, 73–5, 77–9, 81, 83, 89, 96, 101, 104, 108, 118, 122–3, 135, 141, 145, 150
evangelical left 63–4
events 3, 7, 9–11, 14, 34, 48, 69, 72, 79–80, 85, 95, 98, 117–18, 132, 139, 145, 156–7
extinction 1–2, 158–9
Extinction Rebellion 158–9
extremism 13
extremist 6, 15

INDEX

Facebook 46, 48, 50–1, 55
fake news 12, 27, 51
false prophet 79–80, 92, 97–8, 100
False Religion 77, 80, 92, 97–100
Fauci, Anthony 31
FeedSpot 55–6
Feindbild 11–12, 14, 32, 76, 95, 101–2, 129, 138, 144, 152
Feindbilder 11, 17, 77, 87, 92, 101, 143, 153, 156
Fido (film) 10
flatology 133
Foucault, Michel 155
    apparatus 25–6, 31, 77, 111
    knowledge, hierarchy of 26–7, 47
    knowledges, insurrection of 32, 157
    truth, regime of 26, 47, 105, 108–9, 113, 116, 140–2, 153, 157
Fourth Assessment Report 36
free speech 51
Fuller Theological Seminary 63
fundamentalism 72
fundamentalist 59, 73, 96

Gates, Bill 11, 13, 156
gender 29, 36, 41, 152
Genesis 67, 123–4
geographies of knowledge 15–17, 24, 43–7, 49
geopolitical 3, 6–7, 15, 18, 23, 35–6, 44, 72, 82, 85–6, 92, 106, 137–43, 145, 148, 150–4, 156–7
geopolitical imaginations 6–7, 15, 137, 139, 143, 153
Germany 51
global government 3, 76, 93, 111, 137–9, 141
globalist 6, 12, 14–15, 18, 33, 42, 55, 57, 77–8, 81–2, 92–3, 95, 100, 106–7, 110–12, 115–17, 121, 137–42, 145–55
global knowledge 29, 33–4, 36–7
global warming 1, 3, 32, 35, 56, 66, 68, 87–9, 99–100, 106, 110–11, 113–16, 121–3, 126–35, 137, 140, 153, 158, 161
global way of knowing 37

God's chosen nation 18, 145, 148–9, 153
Goebbels, Joseph 111
Gog and Magog 146, 148
Golden Age 69–70
Goldman, Alvin 24
Gore, Al 1, 12, 94, 113–14, 129, 151, 156, 158
governmentality 50
Great Depression 95
Great Reset 7, 78–9, 94, 141, 144, 156
green evangelicalism 66
greenhouse 27, 34–5, 38–9, 56, 66, 86, 89, 94, 99, 129, 160
greenhouse gases 27, 34–5, 38–9, 56, 66, 86, 89, 94, 99, 129, 160
Green New Deal 94–7, 151, 156–7
Gribben, Crawford 59, 62, 69–72, 76, 97, 139, 142, 146, 148

Hagee, John 78, 98, 101
Hal Lindsey Report, The 107
hate speech 50
heaven 69–72, 79, 137, 146, 159
Hescox, Mitch 68
Hitler, Adolf 96
hoax 3, 13, 15, 32, 78, 86, 105–6, 108, 110, 113, 121, 140–2, 151, 161
hoax, climate 3, 13, 15, 32, 78, 86, 105–6, 108, 110, 113, 121, 140–2, 151, 161
Hofstadter, Richard 8, 11, 77, 97, 107, 144, 156
Holy Land 72
homeopathy 54
Homophobia 14, 65

Illuminati 11, 78
immigrants 50, 63
improvisational millennialism 8–9, 15, 31, 54, 156
Index
Infowars 47–8, 50–2, 55–7
Inhofe, James 32
Instagram 51
intergovernmental 4, 21, 30, 36, 41, 76, 94, 137–9, 142, 146

Internalism 28
IPCC 2, 4, 16, 21–2, 28, 30, 33–42, 67–8, 90, 109, 117, 121, 126–7, 132, 135, 155, 159
Islam 63
Islamophobia 14, 65
Israel 72, 98
Israeli 72, 85

January 6 4
Jewish cabals 78
Jews 72
Jones, Alex 48, 50, 52, 78
Judaism 5

Kidd, Thomas S. 61
knowledge-claims 4, 9, 16, 18, 21–2, 24–35, 37, 41–2, 44, 47–8, 51, 53–4, 57, 65, 67, 69, 74–6, 79, 81–3, 89–91, 106–7, 109, 112, 118, 128–30, 134, 142, 157, 160
knowledge hierarchies 13, 16–17, 22–3, 25, 27, 31, 33, 36, 47, 81, 106–7
Kuhn, Thomas 38

LaHaye, Tim 139, 142
Lamb & Lion Ministries 94, 100, 124–5, 129
Lamplighter 100, 106, 129–30
Latour, Bruno 109
Laudato Si' 100, 156
League of Nations 76, 139
Left Behind 76, 142
LGTBQ+ 111
liberalism 92–3, 96–7, 101, 118
libertarianism 93
Lindsey, Hal 55, 72, 77, 85–7, 100–1, 107, 112, 117, 119, 121, 123, 126, 129, 139, 141–3, 149
Livingstone, David 15–16, 24, 30, 41, 43–6, 54, 65–6, 83, 95, 128

McCain John 98
Mad Max 2
MAGA 13
Malthusian 2
Manichean 18, 77, 89, 102, 107, 138, 146, 150, 154

Markey, Ed 94
Marx, Karl
    Marxism 93, 144
mass extinction 1–2, 158
media platforms 6, 47, 49–51
Messiah 3
Meta 46
millennial conspiracism 10–11, 15, 29, 156
millennialism 5, 8–10, 15, 31, 54, 69, 73, 75–6, 101, 107, 138, 146, 154, 156
Millennial Kingdom 72, 79
misinformation 47, 51, 104
misogyny 14, 65
more-than-human 49
more-than-real 46
motivated reasoning 83, 134–5
Muslim 50
Mussolini, Benito 96

NASA 112, 134–5
National Association of Evangelicals 60
natural disasters 1, 35, 40, 71, 88, 129, 158
Nazi 111
New Age 8–9, 98
New Deal 94–7, 151, 156–7
New Left 64
new media 6
New World Order 8–9, 76, 81, 87, 98, 137, 139, 144, 151, 156, 162
Nietzsche, Friedrich 82
Noah 124, 126
normal science 38–40, 90, 130
NOW 7, 56, 78–81, 127, 139, 141–3, 147–51, 156, 162

Obama, Barack 77, 151
Ocasio-Cortez, Alexandria 94
occultism 8
one-world government 3–4, 18, 55, 76, 78–9, 86, 98, 137, 139, 143–5, 150, 153, 156
One World Religion 98

Pachauri, Rajendra 36
pandemic 2–3, 11, 43, 81, 95

Panel on Climate Change   4, 21
Paris Climate Agreement   18, 33, 42, 55, 68, 81, 91, 119, 138, 146–7, 149, 151, 153, 156–7
Pentecostalism   5
Petition Project   128
plague   3, 99
political conservatism   3, 89
Pope   77, 92, 99–101, 118, 156
Pope Francis   92, 99–101
populism   21, 63, 148, 162
populist   2, 12, 32, 48, 62, 156, 159, 161
post-apocalyptic   78
postmillennialism   69–70
post-normal science   38–40, 90, 130
post-truth   21
power-knowledge   13, 16–17, 22, 27, 31, 33, 35, 42, 81, 87, 115
premillennialism   5, 17, 69, 71, 73–4, 76, 78–80, 83, 85, 88, 91, 96, 107, 141, 149, 152–3, 157
Project 2025   162
propaganda   18, 87, 104–5, 108–15, 117, 126, 128, 135, 144
Prophecy   3, 9–10, 50, 53–6, 72, 76, 80, 82, 85–8, 94, 99, 101–2, 106, 108, 110, 117–18, 121, 124–5, 129, 139–41, 143, 145, 148–9, 153, 159, 161
protection   3, 13, 15, 64, 66, 70–1, 86, 88, 90, 93, 97–8, 100, 106, 115–16, 119, 129, 137–8, 141–5, 149–53, 155, 157, 160, 162

QAnon   13–14, 32, 48, 50, 78, 148, 154
queer   14

racist   4
radical Islam   63
radical politics   8
Rapture   5, 56, 71–3, 78–80, 85–6, 88, 91–2, 99, 104, 119
RaptureForums   94, 99, 144, 148, 151–2
Rapture Index   99
RaptureReady   55–6, 93, 98–9, 106, 110–13, 127, 130–3, 141, 143, 147–8, 151, 160

Reddit   51
religious apocalypticism   4, 53, 74, 137, 158
representations of   2, 35
Republican   15, 19, 32, 61–3, 68, 75, 94, 161
Republican Party   15, 61–3, 75, 161
resource scarcities   2
Revelation   1, 9–10, 12, 73, 75, 85, 104–5, 109–10, 140–1, 147, 158
Revelation, Book of   1, 9–10, 12, 73, 75, 85, 104–5, 109–10, 140–1, 147, 158
rightward   2, 4–5
right-wing   6, 13, 21, 59, 63, 78, 81, 102, 109, 148, 162
Robertson, Pat   78, 139
Roman Empire   76–7, 96, 101, 143
Roosevelt, Franklin D.   77, 94–6
Rothschild Family   11, 78
Russia   143, 148, 152
Russian interference   47

Sanders, Bernie   12, 151
Satan   11–12, 73, 75, 86, 91, 98, 102, 107, 112, 128, 140, 142, 147, 149
SDG   143–4
sea level   2, 35, 39, 126
Second Coming   5, 71–2, 78, 88, 91–2, 100, 139, 144–5
secular apocalyptic   3, 10, 156–7
secular conspiracist   4, 6, 95, 118
Shapin, Steven   24
Shaun of the Dead   10
skepticism   14, 17, 26, 35, 67–8, 71, 87–91, 93, 118, 124, 128, 131, 134, 145, 148, 150, 161–2
Snyder, Michael   56, 142
Social Epistemology   16, 19, 21–5, 27, 29, 31, 33, 35, 37, 39, 41, 44, 47, 108, 117
socialist   77, 89, 93, 97, 144, 147, 151
social media   15, 47, 49–52, 55
Soros, George   11
sovereignty   4, 77, 89, 91, 93, 139, 149–51
Stalin, Joesph   96
stewardship   66
St. Francis of Assisi   100
stigmatized knowledge   31

Strandberg, Todd 99, 148
Sugar Land Bible Church 123, 129
sunbelt 14
superconspiracy 8, 16, 54, 75, 78–81, 87, 90, 95, 100, 105, 110–12, 115–16, 119, 121, 125, 138, 143, 156–7
surveillance 3, 50, 137
Sutton, Matthew 60
syncretism 9

technocrats 26
televangelist 78
The Late Great Planet Earth (book) 72
theology 2, 6–7, 63, 65–7, 69, 71, 73, 82–3, 122, 139
Thunberg, Greta 1, 99, 158
totalitarian 96, 141
transformation to the worse 3
transgender 50
Tribulation 71–3, 78–80, 86, 88, 92, 99, 139, 141
Trump, Donald 15, 21, 32, 52, 61, 64, 82, 138, 146, 153, 161
 chief, denier in 32
 conspiracism, presidential 32
 Trump, President 75, 152
Trump presidential campaign 18
truth-claim 14, 16–17, 22–3, 25–8, 30, 34, 57, 61, 69, 83, 92, 95, 102, 104, 110, 115, 124, 133, 139, 142, 155–6
TRUTH Social 52
Twitter 46, 48–52, 55

UNFCCC 36
Ungurean, Geri 93, 127, 141, 143
United Nations 12, 33, 36, 76, 86, 100, 110, 137–44
United Nations Environment Programme (UNEP) 33
United States 3–4, 14–15, 18, 30, 36, 47, 49, 51, 57, 59–60, 62–4, 77, 81–2, 88, 93–4, 96–7, 100, 102, 110, 126, 138, 145–8, 150–5, 157, 162
United States Geological Survey (USGS) 126, 135
USSR 143, 148

vaccination 13, 57, 81, 156
Vatican 100–1
viking 128

Watson, Robert 36
ways of knowing 13–14, 21–2, 25, 42, 44–5, 49, 55, 132, 155, 157
Weber, Max 22, 96
White, Lynn 65
white American 2, 25, 60, 63–4, 81
white apocalyptic evangelical 5, 146
white Christian nationalism 4
white evangelical 14–15, 61, 63–4
WHO 2–3, 5, 7, 11, 13–14, 18, 23, 27–9, 31–2, 36, 42, 45, 48, 50, 52–3, 56–7, 60–1, 63, 68, 70, 73–5, 77–8, 82, 87, 89–90, 94, 97–9, 101–2, 105–7, 109, 112, 114, 116, 122, 124–5, 127–9, 132–3, 139–40, 146–9, 151, 153, 156, 158, 160–1
World Economic Forum (WEF) 1, 158
world government 3–4, 18, 55, 76–9, 86, 98, 100–1, 114–15, 119, 137, 139–45, 150, 153, 156
World Meteorological Organization (WMO) 33
World Wars 72, 79, 85
Worthen, Molly 60

X 46, 50–2, 60

YouTube 46, 49–51, 55, 91, 112–13, 123, 147

ZeroHedge 127–8
zombies 10